SATOSHI SAN

HYMN OF MODERNITY

MACHINE LEARNING, AUGMENTED REALITY,
BIG DATA, QUBIT, NEURALINK AND
ALL OTHER IMPORTANT VOCABULARY
IT'S TIME TO KNOW.

SATOSHI SAN

CONTENTS

SATOSHI SAN

AGE OF
ARTIFICIAL INTELLIGENCE

ARTIFICIAL INTELLIGENCE

Artificial intelligence (AI) is a type of computer software that gives computers the ability to simulate intelligent behaviour. Surprisingly, simulated "intelligence" is not necessarily human-like when it comes to resolving tasks; its potential is even greater.

To illustrate, AI defeated the Chess and Go world champions and has since rendered humans hopeless in future competition. The world champion of Go, Lee Sedol, was defeated by Google's Deep Mind in front of 200 million spectators in 2015, marking the beginning of AI technology race between US and China. Some of the strategic moves played by AI have never been seen in the history of these games but proved to be extremely effective and are still studied.

"AI is more profound than ... electricity or fire"
- Sundar Pichai, CEO of Google

At its current speed of development, AI will soon outpace human performance in a variety of tasks, thus making it economically unviable for humans to execute jobs, resulting in significant automation and economic shifts. AI spending is expected to increase from today's $20 Billion to $120 Billion by 2025.

MACHINE LEARNING

Machine learning (**ML)** is the most powerful and promising artificial intelligence technique to date. The term is most frequently associated with AI based on Artificial Neural Networks technology, which closely resembles the workings of the human brain. The recent ML boom kicked off in the early 2010s, when ML based algorithms blew every other algorithm out of the water during the worldwide visual recognition challenge. Google, Facebook, and Amazon quickly hired top talents from ML research. Today they infuse ML based algorithms in all of their services and operations, claiming they have changed their business models to be AI-first companies.

"Much of what we do with machine learning happens beneath the surface. Machine learning drives our algorithms for demand forecasting, product search ranking, product and deals recommendations, merchandising placements, fraud detection, translations, and much more. Though less visible, much of the impact of machine learning will be of this type — quietly but meaningfully improving core operations."

–Jeff Bezos, Founder and CEO of Amazon

ARTIFICIAL NEURAL NETWORKS

Artificial Neural Networks (**ANNs** or **NNs**) are computerized simulations of biological neurons, like those in the human brain. Inspired by discoveries in early neuroscience, ANNs were invented in the 1940s, but unfortunately didn't show any signs of meaningful intelligent performance at the time. Fast-forward to today, and ANNs are legitimate citizens in nearly every aspect of the digital economy.

Neural networks are very simple in their basic structure, yet capable of displaying astonishing learning capabilities in a wide variety of seemingly unrelated tasks.

As you may know, humans perceive the world through our sensory system, often with our five basic senses. Those perceptions are transformed into tiny electrical signals that travel to our brains to deliver a message. Once reaching the brain, the signals enter a labyrinth of neurons, and continue travelling until the parts that link neurons together, which are called synapses, begin turning off signals that are judged unimportant, while amplifying the ones deemed essential. As a result, signals travel back to a particular sensing organ, carrying a message such as "move faster", "hold very gently", or "that colour matches my sweater"!

Artificial neural networks are commonly referred to as Black Boxes, which is one of their biggest disadvantages. This is because due to their nature, NNs have little to no transparency in the way they make decisions. On one hand this can be seen as a negative feature, but on the other, NNs' complex structure gives them the ability to capture complex information, hence displaying a critical aspect of intelligence. This type of critical thinking is the same reason we are still unable to explain how humans make instant decisions. We may see their input and output, but we don't always know their internal process.

ML TRAINING

Machine learning training consists of applying a standard set of AI procedures. These procedures provide a system for AI to automatically learn from experience in order to perform a task correctly. Performance assessment is generally done by measuring the deviation between the intended and obtained result. The training goal is to reduce deviation to a minimum.

Just like a university student, ML shall manage to learn the expected knowledge through continued trial and correction. Once AI software indicates good results in the training set, it will then move to the test set. AI is often tested on similar training tasks, but not *exactly* the same, just like in school. This is crucial in both making sure that AI manages to generalize its knowledge and in preventing hard wiring. It also allows for some flexibility, which is important in being able to deal with potential deviations in tasks from the ones it was trained on. If the test performance is deemed successful, AI models will be qualified as "working" and can then be used in real world applications.

For example, in the case of autonomous cars, if the

driven mileage without accident surpasses the average human score, then the car will be approved to hit the roads. Intriguingly, this is already the case in certain areas of California.

Types of ML training vary widely and range from supervised learning to reinforcement learning, but generally all remain centred around the basic concept of statistical learning. This means observing multiple series of example data, drawing meaningful hypotheses, and then testing their validity in practice.

There is only one condition for learning, that the experience data or the training environment are not completely chaotic. There must be some degree of natural order, otherwise learning process is not possible, neither for human nor AI. However, AI can sometimes see order where we humans simply don't have enough imaginative power. AI has no physical limits for its IQ; it can be improving to infinity. However, be assured, your brain can also be connected to a computer with large neural networks and you could become as smart as AI. It will be like wearing the much-needed glasses for your blurry vision: Everything will become clear.

SUPERVISED LEARNING

Supervised learning (**SL**) is a type of AI training where a high volume of example data must be provided up-front, from which AI is tasked to extract knowledge. This process is achieved by AI's memory wiring, which links inputs and outputs. AI's memory is a bundle of interconnected mathematical matrices, capable of reducing complex imagery scenes to a few predefined number of categories, known as classes. Examples of classes can include simple groupings such as "good" or "bad" or "cat" or "John" or "ex-girlfriend" or "financial crisis" or "green light" or "ripe apple" or "spoiled apple" or "bigger house than mine".

REINFORCEMENT LEARNING

Reinforcement learning (RL) is one of the most advanced training techniques in machine learning and is said to be the closest viable way of achieving artificial general intelligence.

Contrary to supervised learning, reinforcement learning doesn't need a teacher, previously annotated training data. Instead, RL learns through a simulation program.

Every reinforcement learning problem consists of three elements: a learning agent (the AI), an environment in which the agent learns, and the feedback loop of predefined rewards and punishments that are defined by the creator. Just like in Nintendo video games, Mario receives points when collecting coins or finishing a level. This reward system closely resembles the human dopamine system.

An AI agent has an internal memory state, and after a long series of trials and errors, it should understand that some sequences of action earn a positive reward, some don't impact the agent at all, and other actions penalize the agent with negative points called punishments. The goal of an agent is to learn a behaviour which will maximize his total score.

Every training starts as random agitation. Agent moves left and right and keeps doing so until it receives a reward or gets punished. Moves that led to the rewards get memorized as positive and the others as negative. Frequently incurring positive behaviour gets reinforced, hence the name, reinforcement learning.

The training environment in which the agent learns can be either on real ground or in a simulation replicating physical events in a digital format. Most frequently, agents are trained in simulators because training on real ground is much slower and vastly more expensive, as it becomes necessary to invest in hardware which will likely be destroyed numerous times before the agent learns anything. Training in a virtual simulator can speed up learning a million times over. Only after achieving success in the simulation will the agent need to readjust to a real environment. The better the simulator, the smoother the readjustment period.

Good simulators and powerful computers allow for countless hours of practice in an accelerated manner. In other words, reinforcement learning can potentially provide enough practice for an AI soldier to master every combat sport on Earth in less than a day.

ENVIRONMENTS

Environment is the virtual or real space in which AI agent learns to operate. Environments differ vastly in complexity, making some tasks much harder to master than others. **Environments** can be

- Static or dynamic

- Partially observable or fully observable

- Stochastic or deterministic

- Benign or adversarial.

Static environment means that nothing around you is moving, or is moving very slowly if at all, while dynamic means things are moving and don't stay still, like an ocean.

Stochastic means that the events have a property of behaving randomly, which means impossible to predict with certainty, like a roulette wheel. On the other side deterministic means everything works like a clock, you get what you see, and you can predict to a high certainty the sequence of events, like in a poorly scripted movie with no plot twist.

Partially observable means you are not capable of

seeing the entire environment at each point in time. Fully observable means agent always has God's like ability to see everything.

Benign means you don't need to wonder what other agents inside the environment plot, communication is open and transparent. Adversarial environment means you have an opponent, enemy, or predator. In this case modelling your enemy's way of thinking becomes important, enemy can attack, lie and trick you! Adversarial environments are a great tool to train Artificial Intelligence for self-preservation and aggression skills.

The game of chess is an example of a static, fully observable, deterministic, and adversarial environment. It is static because once the move is made, nothing changes. You can take a snapshot of your current position and evaluate your options, expecting nothing to change in the meantime.

A highway is a dynamic environment, partially observable, deterministic, and benign. It can be considered benign because, hopefully, we don't exist inside of the video game Grand Theft Auto and other cars aren't on a mission to intentionally run us over.

The stock market is an environment which is dynamic, partially observable, stochastic and adversarial – in other words, among the most difficult type of environments to navigate.

EXPLORE vs EXPLOIT DILEMMA

Explore - Exploit dilemma is a trade-off that RL agent must do at each step. He must choose to either use his already acquired knowledge or take an opportunity to learn new knowledge by experimenting.

When a RL agent begins to learn a new task, he starts with no assumptions about the world. He has no prior knowledge on which actions lead to rewards and which lead to punishments, so he starts by taking completely random actions and memorizes the results. From an outside perspective it seems like random agitations of an indecisive drunk person. However, after a given period of time, the agent has collected different state-action results and is now faced with a choice: shall it take the best-known action from this moment moving forward? Or should it continue to experiment and agitate in hopes of discovering some better strategy? This dilemma is called the *Explore – Exploit Dilemma*, and finding the right exploratory ratio is crucial for correctly training AI agent, especially if the training time is limited.

As a rule of thumb, this ratio starts at 50%, which means the agent tries new actions half of the time.

Then the ratio starts to gradually decrease until stabilizing at around 10% for most environments. This is the stage where the agent becomes *greedy*; it has seen the world and is now set to collect the most rewards possible by using his best-known strategies. However, as you noticed, the ratio is not completely null, and agent keeps trying something new 1 out of every 10 actions. This is important as it will ensure that agent stays up to date with potential changes in the underlying environment. Environments often evolve and the reward structure from a day ago might have changed since, hence the agent must adapt.

If the environment is known to be static and benign, then the ratio will usually descend to 0%, meaning the agent has achieved optimality and will no longer explore.

DELAYED REWARD MECHANISM

During a training process, an AI agent may discover that not all rewards are created equal and that some sequences of actions may lead to greater benefits. But as in real life, one will usually need to invest significant time and effort upfront, refusing instant gratifications, with the aim of achieving a much bigger reward later. AI agents can discover such strategies as **delayed rewards** through a diligent environment exploration process.

The delayed reward mechanism makes an integral part of our daily lives, and those who master it well are often more successful than those who don't.

Should I spend five years in college or start earning right after school? Should I buy this shiny new car, or should I invest in stocks that pay dividends instead? Should I cheat on my wife or not? Should I punch his face for insulting me or should I just walk away?

As children we learn to make trade-offs and are supposed to be smarter about those as we become adults. AI agent too, learns to evaluate his options and avoid falling into the trap of immediate benefit if such will hurt the chance of a higher future return.

OVERTRAINING

Also referred to as over-fitting, **overtraining** is one of the most common beginner mistakes when training AI. Overtraining results in poor AI performance. Problems come from the counter-intuitive belief that spending more hours on training will always yield better results. The reality is surprisingly different and is often due to the limited size of the training set. When passing a predetermined optimal training time over the same training set, AI learns too many details. Although this may seem counter-intuitive, it is in fact harmful.

To suggest that a margin of surprise shall be left inside an AI algorithm is a very philosophical idea. Allowing room for open-mindedness correlates with the capacity to generalize, which consequently increases intelligence in the long term. Achieving the proper balance between historical experience and open mindedness about new possibilities is the apex of critical thinking.

Luckily AI engineers have designed parameters that indicate when AI is entering an overtraining phase. For example, it becomes evident that AI is overstrained when it performs very well on the training tasks but performs poorly on the testing tasks. In general, reducing the time of training will increase the testing score, all the way up until the point where we reach the undertraining zone. As a result, the smartest plan of action is to continue progressing with training as the testing scores increase and stop as soon as performance becomes compromised.

ADVERSARIAL TRAINING

"If you know the enemy and know yourself, you need not fear the result of a hundred battles."

–Sun Tzu

Adversarial training is a powerful type of training which involves two or more AIs competing against each other. This is important, because contrary to popular belief, AI is not a mathematician. If you look under the hood of the AI as it is learning to play chess, you'll see that it's not calculating all of the possible plays and picking the best ones, this is simply impossible due to the combinatorial explosion of possibilities. Even the fastest supercomputer would take millions of years to complete this task.

Instead, AI is very much like a human, in that it is a statistical learner and takes a snapshot of what is currently happening, calculates a few possible actions, and chooses the best one based on past experience. This makes it hard for AI to learn an adversarial game without an opponent, the same way it is difficult to play chess alone. Bringing in an adversary which battles against you with methods and techniques you're not formerly aware of is an excellent way to expand your knowledge and broaden your skills.

DEEP LEARNING

Deep learning (DL) is the most advanced set of machine learning functions, consisting of an increasingly large number of interconnected artificial neural networks. The additional layers allow for the capturing of more information, resulting in a more knowledgeable AI. The higher the number of layers, the "deeper" the network and the deeper the expertise.

"Artificial Intelligence, deep learning, machine learning—whatever you're doing if you don't understand it—learn it. Because otherwise you're going to be a dinosaur within 3 years."

–Mark Cuban, Billionaire

BIG DATA

All Google searches, e-mails, Instagram posts, WhatsApp messages, Tweets, OneDrive files, iCloud pictures, Netflix movies , population tax data, people's genome data – all of that staggering amount of information being produced every day, by every connected device all over the globe, is **BIG DATA**.

"Big Data will spell the death of customer segmentation and force the marketer to understand each customer as an individual within 18 months or risk being left in the dust."

–Ginni Rometty, CEO, IBM

Data has always existed; accountants have always kept records of city financials, scientists have kept track of experimental observations, and authors have published newspapers and books documenting current events. But big data only became possible with the introduction of revolutionary technology that allowed for inexpensive storage and processing of substantial amounts of information.

Most important technological innovation igniting the big data revolution was Google's 2004 publication

Map Reduce, which aimed at connecting many low-cost computers, effectively distributing work between each node. The core business of most of today's internet giants like Google, Amazon, Facebook, Netflix, and others relies on successful handling of big data. Consequently, the first two also became major cloud providers, essentially renting storage space on their computers equipped with this technology to any business worldwide, and for a low price at that.

To emphasize the importance of big data in the current AI boom, consider this: The first idea of AI, called "automata", was introduced by the Greek philosopher Aristotle in 300 B.C. Centuries later, during the Renaissance, Leonardo Da Vinci created multiple complex engineering machines capable of operating by themselves. In the 1940s, English mathematician Alan Turing built on the works of the Greeks and pushed Artificial Intelligence Theory to definitions accepted worldwide. Turing's invention of the universal machine, which was considered the first viable computer, would result in an unprecedented inflow of investment into the field.

At this point, there appeared to be everything required for AI to finally emerge: incredible calculating power provided by IBM, powerful algorithms coming from neuroscience research, inexpensive electricity, two thousand years of compounded mechanical engineering knowledge, astonishing manufacturing facilities, and yet, the only Terminator walking the Earth was an Austrian-born bodybuilder. What was lacking? To everyone's surprise – data! Emergence of big data is considered as perhaps the final essential ingredient for building AI – from there it is only a matter of iterations.

Today Big Data sits within the Business Intelligence and Analytics department of every major corporation. And although Generally Accepted Accounting Practices (GAAP) do not yet permit data to be capitalized on the balance sheet, we know that data is extremely valuable because five of the world's seven most valued companies are essentially data drive companies.

RESPONSIBLE AI

"If you're not concerned about AI safety, you should be. Vastly more risk than North Korea."

–Elon Musk

Responsible AI is a set of ethical guidelines and techniques used to audit artificial intelligence algorithms in order to prevent bias, malfunctioning, or evil intentions. This applies especially to AI algorithms used in organizations that make decisions regarding humans, such as court judgments, bank loan applications, and HR departments. Responsible AI is achieved by ensuring that the training data is both well-selected and unbiased and AI decisions are audited for transparency and explainability.

NARROW AI

Narrow AI is a type of AI that can only resolve a single task, but usually with the benefit of high performance. The Holy Grail of AI is Artificial General Intelligence, where one single artificial intellect will learn to drive a car, speak Spanish, advice on Law, salsa dance, and make killer mojitos. However, creating such an AI is extremely difficult and we remain far from it. As long as AI is able to perform tasks faster, cheaper, and better than humans, we don't have to wait for AGI in order to begin reaping economic benefits.

NATURAL LANGUAGE PROCESSING

Natural language processing (**NLP**) is a mixed field consisting of linguistics, computer science, and increasingly AI, and focuses on working with language related tasks. Natural language processing is used by businesses for text sentiment analysis, text summarization, chatbot user interface, text translation, text transcribing, and even text generation.

TURING TEST

The **Turing Test** is a test aimed at qualifying the level of machine intelligence. The test was introduced by Alan Turing in his 1950 paper, *Computing Machinery and Intelligence*. His paper opens with the words: "I propose to consider the question, can machines think?" Because "thinking" is difficult to define, Turing chose to replace the question and proposed an "imitation game" – if the computer managed to successfully convince the interrogator of its aliveness by appearing human, then the machine passed the test.

To conduct the test, a human subject talks to two partners who are either hidden behind curtains or set to interact via a chatroom. One subject is a machine pretending to be a human and the other is an actual human being. Both the machine and human compete in trying to fool the human judge to correctly differentiate man from machine. If the machine manages to trick the judge, the test is passed.

We are very quickly approaching an AI with language abilities powerful enough to pass the test, which means we are indeed in very exciting times. Once AI has surpassed this point, a customer call centre can become fully automated, and many other similar types of jobs will follow suit.

COMPUTER VISION

This branch of computer science is preoccupied with understanding how a computer can process visual media content, such as an image or a video. A deep learning-based AI, known as Convolutional Neural Network, launched a revolution in the field of **computer vision** in 2017 by beating human-level accuracy in object recognition tasks.

One of the first fields to bet big on computer vision software was the postal service. Much of the research was conducted to create optical character recognition software for manually written addresses, which would greatly automate and speed-up letter and package routing. Today, computer vision applications are numerous, with facial recognition entering the "hype zone", because of Big Brother initiatives like China's social scorecard program.

CLASSIFICATION

Classification is a supervised machine learning task aimed at sorting input data into predefined categories. Classification is one of the most abundant and frequent tasks the human brain accomplishes at all times.

Today, a deep learning model can learn to classify almost anything as accurately as a human. Filtering spam emails is one of the most common classification task examples. In fact, Google, Baidu, and Yandex employ very sophisticated AI algorithms to sort out spam while preserving legitimate emails.

Another lucrative domain for AI classification task algorithms is medical imagery analysis. AIs can diagnose a patient based on their medical imaging studies *better than a professional doctor* at a fraction of the price.

FACIAL RECOGNITION

Facial recognition (**FR**) involves AI that is capable of providing the identities of people present in a photo or video. Facial recognition can be used by a specific company as a security measure to prevent unwanted intruders on their premises, or by an entire nation as a means to oversee citizen behaviour. For example, in China FR is employed to prevent mass gatherings, control demonstrations, and monitor crime.

DECISION TREE

Before the deep learning revolution, decision trees were among the most successful techniques for re-solving classification tasks. They remain widely used for their transparency and reliability.

To classify data between a given set, **decision trees** find particular differences in the data which allows for maximal class segregation. Next, decision trees branch out, further splitting into two classes with the process continuing until each data point has been classified. The endpoints of each branch are called leaves. If you have ever seen a tree of life model of biological evolution, you are familiar with the most basic branch split between eukaryotes and bacteria, then the split between mammals and reptiles, until eventually you come to the human, which is consid-ered a leaf. This a basic decision tree structure.

RANDOM FOREST

Random forest is an ensemble learning method based on decision trees.

In a nutshell, as data grew and became more complex, it was no longer enough to design a single decision tree to achieve maximum accuracy in a classification task. As a result, computer scientists developed methods of producing anywhere from dozens to hundreds of decision trees with different settings, hoping that each will pick up something special about the data. After that, each decision tree provides an opinion on how it would classify the content that was asked of it, votes are counted, and at the end the most common opinion is established as the correct answer. Random forest is a very simple and democratic form of AI.

REGRESSION

Regression undermines a certain property of an environment where one or multiple variables can explain evolution and behaviour of another variable of interest. Regression learning is possible only in environments that display some degree of order and can be statistically studied.

For example, a simple regression can model the trajectory of a cannon ball. A more complex regression can make a price estimator for real estate based on input properties like location, size, number of windows, and year of construction. It is a supervised learning task so AI first needs example data to train, but that is usually not an issue. Machine Learning algorithms are great at building regression models even for the most complex environments.

Science, insurance, and supply chain are among the fields that are heavily relying on regression models.

SENTIMENT ANALYSIS

Also known as sentiment mining or emotional AI, sentiment analysis (**SA**) refers to the use of AI for quantifying and qualifying the emotional charge of news, tweets, financial reports, customer reviews and camera footages. Contrary to the common belief that AI doesn't recognize common human emotions such as rage, fear, concern, happiness, or essentially any human emotion – recognizing human emotion is actually one of the simplest tasks AI can learn to do.

Today we count 27 official distinct emotions: Adoration, Aesthetic Appreciation, Amusement, Anxiety, Awe, Awkwardness, Boredom, Calmness, Confusion, Craving, Disgust, Empathetic pain, Entrancement, Envy, Excitement, Fear, Horror, Interest, Joy, Nostalgia, Romance, Sadness, Satisfaction, Sexual desire, Sympathy

As if preferred by evolution, our facial expressions and language are extremely expressive and immediately reveal our emotions, making it an easy job for AI to capture. It is actually more challenging to hide emotions from AI rather than communicating them. Confirming the theory, an AI capable of analysing facial expressions recently defeated world poker champions.

GENERATIVE ADVERSARIAL NETWORKS

Generative adversarial networks (**GANs**) is a branch of deep learning that focuses on generating new media content. Media that GANs can create include paintings, music, cartoons, videos, movie scripts, and more. GANs are designed with two different networks that compete with each other – one designs content and the other criticizes its quality. How does the critical network know what humans will consider good or bad? We train it for that exact purpose, but only after giving it the content of the generative part.

SATOSHI SAN

DEEP FAKES

Deep fakes are fake news, fake images, or fake realistic videos created by AI. They are called fake because their primary goal is to mislead the audience into believing they are real. An AI deep-fake-generated video of Barack Obama giving a speech created a massive buzz a few years ago; the video seemed nearly indistinguishable from the authentic Obama. We are not talking about Hollywood graphics or cartoons, but video and sound of such high quality that it is indistinguishable from the real. This opens a breach to extremely misleading news and potential political issues.

Because the web will soon be flooded with deep fakes, it is said that in the future all reputable media providers will need to authenticate their content with a blockchain token to prove the news is not fake.

ROBOTICS AND CONTROL

Robotics is the branch of engineering that deals with designing, programming, manufacturing, and operating robots. Robots are designed to assist humans in day to day living, therefore industries are leveraging robots to automate human work in places where it makes economic sense.

Historically, the field of robotics heavily relied on human programmers to manually code every possible situation in which a robot might find itself. In many real-world situations, such a task is simply impossible. However, Japanese engineers refused to give up and spent almost two decades unsuccessfully sweating over a practically unresolvable task. Progress lagged until a team of Canadian researchers teamed up with Silicon Valley engineers came up with the Deep Learning technology. Deep Learning finally allowed robots' motor intelligence to be wired inside virtual simulators during a training process, with no manual coding needed.

Today, provided that you have the right hardware and a simulator, teaching a robot to walk can be as easy as importing a few Python libraries and writing a few lines of code to add specifications.

The battle for better robots is actively ongoing in the hardware space. This is where good control systems make a world of difference.

Control theory emerged from a combination of mathematics, mechatronics, and computer science in the last century. At its core, control theory studies various ways of stabilizing systems and smoothing out their operations.

In a simple case of a drone, the smarter the embedded control system the more stable it is, making it a piece of cake to fly. On the other side, cheap drones that don't integrate advanced control systems are very hard to fly and you oftentimes end up bashing them into trees or into the ground – they simply don't have the necessary self-stabilizing layer.

Recent advancements in robot control are peaking strong, and you have probably seen videos of Boston Dynamics's Atlas making backflips like a pro. Now that we have a Deep Learning brain which can intelligently command the robot and an advanced control system which allows for a smooth navigation, the only last major piece remaining before we get into the era of massive robot deployment is the lack of light and durable energy source. Unfortunately robots consume a lot of energy and carrying around a heavy battery lasting for an hour tops is not a viable option.

SIMULATION

Simulation can and does exist in a physical environment. But usually it refers to a virtual environment intended to replicate a real-life event in an experimental manner. **Simulations** go far beyond games and represent a powerful tool for virtually experimenting with chemical reactions, physical events, and even human decision-making processes. Economists use simulations to predict markets, bioinformaticians use simulations to accelerate drug discovery, and automakers use simulations to build aerodynamic cars.

"On occasion, we at 'MythBusters' come across stories we want to test that require using a pig carcass to simulate human physiology."

–Jamie Hyneman

Simulations are increasingly used as training environments for artificial intelligence. In the same way you might mentally visualize new judo moves before trying them in the real world, AI agents first train in a virtual environment before testing on the ground.

Modern digital simulators are much better at simulating physics than the human brain, hence providing better results. In fact, simulating real life events using digital software is becoming so accurate that soon the simulation will be *indistinguishable from reality*. This will result in a line of powerful and hyper-knowledgeable AI agents. Imagine if you had a chance to live 1000 lives and you could accumulate every experience from one life to the other. The 1001st version of yourself would be what Rambo is to a small child.

OPTIMIZATION

Optimization is the approach of finding the best possible solution to a problem. As you can imagine, there are infinite ways to approach every problem. Some approaches can lead to a solution, while others have no solution. Solutions can also vary on a scale of performance, such as whether cash is needed to resolve the problem or if time would be the ultimate resource. Usually, the fastest and cheapest solution is considered the winner.

If a problem is easy and exists in a simple environment, then we can typically find "the obvious solution". But, consider the plan of action if a problem is complex, such as delivering every postal parcel in a city in the least amount of time possible. Such a difficult problem truly has millions of solutions – actually, to be precise, tens of trillions of possible solutions. There is no obvious one solution, so it appears you'll have to try them all.

But wait, you don't have time to try them all. So, what do you do? Do you try to see if you can eliminate a big chunk of potential solutions in one swift, easy action? The answer is always 'yes'. You usually attempt one solution, then try to add or remove some small chunk of the solution, replacing it with what seems a better one, and you continue with this pattern until you arrive at a solution where you can no longer improve. When you reach a place where any further additions or subtractions only makes the problem worse, that's how you know you've optimized a problem.

In simple terms, optimization is a technique of settling "in a happy medium" of a difficult situation. As usual, with an optimization problem many various parties seek the biggest piece of the pie and have specific reasons to justify their demands. As you might imagine, it is challenging to quantify the problem with numbers, but that's the exact function of taking an optimization approach in a preliminary step.

Customers normally want to buy cheap; businesses normally want to sell expensive; and unfailingly, both will fight to the death on behalf of their opinion. In the end, nature will optimize.

GENETIC ALGORITHMS

A genetic algorithm is a machine learning technic used for successfully solving optimization tasks. Genetic algorithms are easy to set up and intuitive in their processing, as they are heavily inspired by the concept of Darwinian evolution and natural selection.

Genetic algorithms start by initializing a random set of solutions to a given problem. According to a predefined criterion, the best set of solutions are chosen and become parents. Parent solutions are melted together in various ways and propose the next population of possible solutions. In this process, some of the "children" solutions will receive random tweaks here and there, slightly altering them from their parents, a process which is called mutation. Once a new population of solutions is created, they will again be assessed against the defined criterion and the superior solutions will be selected as parents for the next generation. And as long as each new generation performs better than the previous one, the process of optimization continues. At the point where improvement is no longer seen, we have arrived at an optimum. Note that it is not necessarily the global optimal state, but it is definitely a much better solution compared to the majority of other solutions.

PYTHON

Besides being a huge scary snake from Southeast Asia, Python, as you might know, is also a programming language. Recently, Python language has been buzzing all over the corporate world. Many Excel users are scared they may be replaced by a Python expert. With **Python** you can easily automate tasks, plus Python is the default language for designing and training artificial intelligence.

Python is the C-based programming language developed at the beginning of the 90s and is becoming an increasingly popular tool for all data tasks. Even the biggest investment banks, such as Goldman Sachs and JP Morgan Chase, are actively hiring hundreds of Python programmers to manage their tremendous amounts of data.

Python's main advantage is its relatively light syntax; it is definitely one of the easiest programming languages to learn. In less than a month you can be fully operational, though it takes many years to master.

ANALYTICS

"Customer data in the hands of marketing analytics experts is the most effective weapon of the 21'st century"

-Matviy Kruglov, Customer Sales Executive at Salesforce

Analytics is a field dealing with collecting, processing, and interpreting data in order to extract and communicate useful insights. Analytical tasks can be completed by hand if the material is paper-based. If the data is digitized, it can be completed with the aid of analytics tools such as Excel, SQL, Python, Qlik and Tableau. Analytics are embedded in all business aspects and aim at improving business performance through conclusions drawn from factual data or customer feedback analysis. Analytics teams are often located inside company's Business Intelligence (BI) department.

AGE OF CYBERWARS

PRIVACY

Privacy is a multi-purpose practice shared between politics and computer science designed to ensure that citizens' private data is both safe and anonymous. This is especially important when it comes to highly sensitive personal data, such as medical records, internet history or credit card details.

"I don't like to share my personal life… it wouldn't be personal if I shared it."

— George Clooney, Actor

GDPR

The General Data Protection Regulation (GDPR) is the new law enforcement directive and formal list of rules concerning the protection of customer, employee and citizen data. GDPR was issued by the European Union and implemented on May 25th, 2018, as the newer version of the first Data Protection act issued in 1995.

GDPR was taken extremely seriously from its beginning. The diligence was likely due to the non-compliance fines being set at the steep price of $14 Million or 4% of total revenue, depending on which figure is larger.

Besides giving customers the right to access their own data or ask to be "forgotten", GDPR also regulates automated algorithmic decision-making when it comes to making decisions regarding humans, obliging companies to have people supervisors in the loop. This is especially important to avoid programmed algorithmic biases when it comes to social allocations, job-posting reviews and medical diagnosis. Companies are also enforced to encrypt user data and protect it at all cost.

For illustration purposes here is an official extract from GDPR on the subject of algorithmic decision making:

REGULATION (EU) 2016/679 OF THE EUROPEAN PARLIAMENT AND OF THE COUNCIL

of 27 April 2016 on the protection of natural persons with regard to the processing of personal data and on the free movement of such data, and repealing Directive 95/46/EC (General Data Protection Regulation)

Extract from Clause (71)

"The data subject should have the right not to be subject to a decision, which may include a measure, evaluating personal aspects relating to him or her which is based solely on automated processing and which produces legal effects concerning him or her or similarly significantly affects him or her, such as automatic refusal of an online credit application or e-recruiting practices without any human intervention."

CYBERSECURITY

"Information is the oxygen of the modern age.
It seeps through the walls topped by barbed wire,
It wafts across the electrified borders."

-Ronald Reagan, US president

Cyber security or computer security, sometimes called network security, is a branch of computer science focusing on the privacy and safety of interaction in the cyber space. Cyber security research focuses on testing and designing ways to better secure systems and devices from unwanted intruders. Cyber security consultants focus on advising corporations and governments on the best practices to ensure their company's network is secure and their data is protected.

Computer security historically is organized around three aspects: confidentiality, integrity and availability. On the other side, the cyber-attacks generally aim at infiltrating systems and stealing, altering, or destroying sensitive information. Cyber criminals often have even more threatening agendas in mind, such as extorting money, committing industrial espionage, or creating large-scale propaganda campaigns. And strangely for these online assailants, sometimes it's just for fun.

There has also been an ongoing cyber war on an international scale, with countries like China, the U.S., Iran and Russia topping the lists of cyber scandals. Although, naturally, each is consistent in denying their own involvement. We have seen no shortage of nationwide attacks in recent years, such as the Stuxnet attack, when hackers introduced malicious code into Iranian nuclear facilities, effectively throwing their nuclear development programs years backwards. Or take for example the alleged hacking of the 2016 U.S. presidential election, where Russians were accused of utilizing Artificial Intelligence to artificially boost Donald Trump's campaign in an effort to avoid a Hillary Clinton presidency.

While some cyber-attacks are being done with malicious intent, you may be surprised to learn that many of these online attacks are done by the hands of purely curious teenagers experimenting with manipulating the intricacies of computer networks. A notable example of this is the 2000 Distributed Denial of Service attacks launched by a 15-year-old Canadian boy which targeted a number of influential international commercial websites including Amazon, eBay, CNN and Yahoo. And surprisingly, the DDoS attack worked.

The total bill for the damages amounted to no less than $1.2 billion dollars. The teen was dubbed "Mafia Boy" for his viral transgression and sentenced to eight months in open custody, with his juvenile status helping avoid serious jail time.

Currently, the hottest ongoing topic on the cyber research front is cyber cryptology, which is a purely mathematical problem that can undermine even the most modern internet security. Every important line of internet communication is encrypted so that an intruder who attempts to spy on network communications between a sender and receiver will not manage to read the messages he intercepts unless he has access to the user's password.

However, intruders have become crafty with their trickery, often using tactics like pretending they are your system administrator and requiring you to change your password based on a recent hacking and ask you to click on a certain link. As expected, this fake link will direct you to a fraudulent web page that allows the hacker to capture your security information.

NSA

The National Security Agency is the largest intelligence agency in the United States Department of Defence. Yes, you heard that right. They heard that right too, because of course, they're probably listening right now. Why? Allegedly to protect the innocent taxpayers from the bullies.

The NSA began as a bureaucratic unit tasked with deciphering enemy communications in WWII, which in simpler terms means gathering a bunch of maths enthusiasts and throwing them into the world of foreign relations. Truman made this task-force official in 1952 as the primary agency responsible for monitoring communication of domestic and foreign targets as part of central intelligence.

Today NSA employs not only mathematicians who create the ciphers but also the hackers who test them. NSA also employs hackers to establish surveillance systems inside and outside the country. Some of the hackers are graduates from top computer science universities like Stanford, others are ex-criminal hackers who have been caught but then offered the chance to collaborate, in exchange for a reduced prison sentence.

NSA made big headlines in the late 90s when much of the internet was still considered to be a virtual Wild West. As time went on into the 2000s, the NSA fell out of the public eye up until 2013, when a former NSA contractor Edward Snowden testified in the Supreme Court and revealed entire schemes of spying and secretive surveillance that the NSA was using to listen in on both domestic and foreign officials.

All things down, you and I usually benefit from the NSA more than we care about what they do with our data. Without the NSA the internet would become a dangerous place and would probably disappear. However not all artificial intelligence researchers would agree.

"I am scared that if you make the technology work better, you help the NSA misuse it more. I'd be more worried about that than about autonomous killer robots."

- Geoffrey Hinton, Deep Learning pioneer

PROJECT ECHELON

Project ECHELON was originally a secret government surveillance program created in the 1960s, an operation used by the United States, UK, Australia, Canada and New Zealand to spy on the Soviets.

By the end of the 20th century, ECHELON had evolved beyond its military and diplomatic purposes and grew into a global surveillance tool used by the NSA. Edward Snowden's whistleblowing revealed further details about the massive scale of ECHELON operations.

Britain's *The Guardian* referred to ECHELON as a global network of electronic spy stations that can eavesdrop on telephones, faxes and computers. It can even track bank accounts. This private information is stored in Echelon computers, which has the capacity to hold millions of records on individuals. However, in an official capacity, Echelon doesn't exist.

SOCIAL ENGINEERING

Social engineering is a hacking technique which takes advantage of the weakest chain in modern cybersecurity systems, which of course will always be the individual using it. Social engineering hacking doesn't look for mathematical weaknesses of security algorithms but studies the ebbs and flows in human behaviour and logic. Although a company's firewall might be incredibly secure, all it can take to inject a lethal virus is an employee curious enough to stick a USB he found lying around into his work laptop.

Hackers that become Master of Social Engineering are amazing con artists with strong technical skills and excellent emotional intelligence. They can easily place a call under the guise of being a new pet food campaign, claiming that all you need to do in order to qualify for a free sample is to give your name, address and your dog's name.

These pieces of information, of course, are among the most frequently used security questions used to recover forgotten passwords by many banks and other institutions.

DUMPSTER DIVING

Dumpster diving, or in other words searching for information in the trash bins, is commonly used by hackers and spy agencies to dig for information. It is of course much more applicable for corporate trash picking.

"Oracle for example has even hired people to dumpster dive for information about its competitor, Microsoft. It's not even illegal because trash isn't covered by data secrecy laws."

-Kevin Mitnick, most wanted hacker of the late 90s

NETWORK TRAFFIC ANALYSIS

Traffic Analysis is the study of communication patterns. Without having access to the content of the messages, traffic analysis can still indicate plenty of relevant information.

Who is talking to whom? How long do the conversations last? What time do the conversations take place? Those and all similar questions help to understand the dynamic of the exchange.

Of course, the plain "who talks to who" is already a lot of precious information, but there is more traffic patterns that are well known to the military. For example, regular communication from Alice to Bob and Sam, and then from Bob and Sam to dozens of other individuals, most certainly means Alice is communicating a chain of orders.

Rapid and short communication between Alice and Bob – can denote negotiation.

NETWORK

A group of computers or mobile devices linked to-gether, most commonly through a wire or a Wi-Fi connection. Most local networks are connected to the internet, making them vulnerable to attacks from hackers. If one of the devices connected to the net-work is compromised, then it's safe to say that *all* the devices connected to the network are compromised.

Networks are the most frequent channels for cyberattacks, as a hacker doesn't need to risk physi-cally approaching his target but can simply commit his theft through the internet from the safety of his couch somewhere on the other side of the world. A com-puter hacker may not be able to walk up to a large professional athlete and rob them on the spot, but with just a little bit of personal information they are able to inflict even greater damage.

We typically distinguish three distinctions of net-work groups:

1. Personal home networks, which usually in-cludes all family's computers, mobile devices, and private personal information. This is usu-ally thought to be the safest place to store in-formation.

2. Public networks, which are usually accessed free of charge in cafes, airports and hotels, which often makes them a totally vulnerable place.

3. Corporate network, which of course you access at the office. Company networks are the Eldorado for cyberattacks as these networks store sensitive customer data, industry secrets and financial portfolios. Such a hack could bring even the strongest of companies to the ground, which is why virtually every single business conglomerate invests millions of dollars in cyber protection and security auditing every year.

The internet is one big network made of many small public, private and work networks. Contrary to what you might think, the internet is not a benign environment but an adversarial one. When you go out of your private network onto the internet's public network such as *Cartier.com* or *SomeLuxuryRestaurantReservation.com*, to browse for some content or make an online purchase, a bunch of hackers are always watching the newcomers and are ready to follow them back to their private network in an attempt to spy on them or steal their data.

P2P NETWORK

A series of interconnected computer nodes forming a private network to send each other usually illegal resources over the internet. P2P architecture was originally popularized by Napster, which allowed for pirated music sharing. Today P2P networks such as *Torrent* allow for distribution of illegal copies of movies, music, books, software and other forms of media, causing devastating revenue losses to the media industry. P2P is the concept of a Sunday street market, where transactions are directly undertaken between the seller and the buyer, in contrast to Walmart, for example, who takes the position of a middleman between the two parties.

The idea behind P2P served as the cornerstone for democratization and decentralization of digital services. Instead of bbc.co.uk we got Facebook and Instagram, where everyone can publish and share content. Instead of city taxi we got Uber. Instead of Hilton hotels we got Airbnb. Instead of national TV we got YouTube.

While the idea of pure decentralized P2P network moved in most cases to the Dark Web, its direct

cousin in the form of centralized social sharing platforms became a tremendous worldwide success, effectively creating modern internet giants.

With time however every service finds its own
niche, few Airbnb hosts can rival Hilton's quality and
few YouTubers qualify for an Oscar. Both types of
offers now coexist in a healthy competitive ecosys-
tem.

DARK WEB

Dark web, the Black Market of the internet, is not visible to search engines like Google and requires special software, configurations, and authorization to access. It is hidden from the public to avoid undesired attention, since it conveys mostly illegal services.

"Being able to see an activity log of where a kid has been going on the Internet is a good thing."

–Bill Gates

Services conducted over Dark Web include human and organ trafficking, prostitution, exotic animal trade, weapon trafficking, and of course, drug trafficking. The dark web also proposes a series of even more immoral services such as online showrooms where authorized visitors can assist via web camera acts of torture, violence, pornography and what not, for a fee. One usually settles transactions on the dark web with anonymous Bitcoins, which boosted Bitcoin popularity and led to the Bitcoin bubble of 2018.

TOR

Tor is a software which allows users hide their true IP and browse internet anonymously, thus making it very hard to be spied on. Tor is very popular amongst hackers and criminals as it gives them the desired free-dom to conduct criminal activities over the internet as well as accessing the Dark Web.

HACKER

"There are few sources of energy so powerful as a pro-crastinating college student."

— Paul Graham, Hackers & Painters: Big Ideas from the Computer Age

A hacker is a skilled computer programmer capable of mounting successful cyber-attacks, often by finding bugs and weaknesses in a target network. Most hackers are considered to be criminals, but there is a small subset of good hackers called White Hat Hackers.

Many hackers start young and by the time they hit adolescence they already have a better understanding of computers than most people ever achieve. Ironically, this is because arguably the best way to learn the limitations of a system is by trying to break it.

If it finds itself in the right hands, such knowledge can then be used for good in order to design better systems. One of the best examples of this positive brand of hacking is the co-founder of Apple, Steve Wozniak. In college, Wozniak was in a "young hacker community" which designed a tool that allowed a person to use phone booths to make phone calls for free. Wozniak later leveraged his extensive knowledge of computer systems to help Steve Jobs launch their revolution in personal computing.

A **black hat** hacker, commonly known as an encryption-cracker, is the brand of hacker who uses his advanced skill set for evil. Although in pop culture they are idolized as individuals with superhuman powers, witty intellect, and playful character, in reality these crackers are criminals with an aim of stealing, destroying and damaging. They usually do it based on money, ideology or personal satisfaction.

A **white hat** hacker uses his advanced computer skills for solving problems, having good intentions in mind and operating within the law. White hat hackers are often hired by companies to audit internal vulnerabilities and assess potential security breaches.

Hackers that specifically invited prior to a software launch for bug and security testing are called **blue hat hackers**. This term is extensively used in Microsoft.

MALWARE

Malware stands for malicious software, as it is a type of software written by hackers and transmitted to victims in order to steal information, destroy information, or spam/spy on other computers. Hackers can spread malware via email attachments, websites, USBs and more. There are plenty of different malware categories out there, here below are few of the most prominent ones.

*"A **better** biological virus — like a computer virus — will perhaps just make its host sick, but still well enough to keep spreading the virus."*

— Peter H. Gregory, Computer Viruses for Dummies

Virus is one of the most primitive types of computer malware. A virus begins its life when an executable file to which it is attached is executed on the victim's laptop. Virus code can spread to other user applications and cause mild to severe symptoms like data corruption, malfunctioning or network spamming. Viruses cannot infect other computers on their own and require infected applications to be transferred by the user as an attachment. Of course, this is all done without the user realizing that software is corrupted. Although an antivirus is a great protection against viruses, viruses can avoid antivirus software for long periods of time through digital mutation.

The first virus was created in 1983 by a USC student for pure demonstration purposes. Today there are more than 100,000 different varieties of viruses, some deadlier the others.

A computer **worm**, technically a subclass of a virus, has all of the same properties and functions as a virus with the exception that worms don't need the host to send out infected files in order to propagate. Worms usually hijack the computer's communication systems in order to inseminate on their own as fast as they can, which makes them extremely devastating. For example, they can send themselves to all the emails found in your contact list, do the same to each of the victims and to their victims and so on.

The first worm, created in 1988, was able to single-handedly crash 10% of the early internet.

The third most common kind of computer ailment after viruses and worms, a **Trojan horse** is perhaps amongst the most primitive, as it doesn't replicate or reproduce, and strikes only once the infected program is initiated by the user. Trojan horses can remain undetected for long periods of time before being identified, hence the name.

In case you are not familiar, this is referring to the mythical wooden statue of a horse that Trojan soldiers paraded into their city as a trophy of victory, not realizing there was an onslaught of enemies waiting inside the statue. As soon as the sun set, enemy soldiers broke out of the horse and forcefully opened the city gates, leading to the swift demise of the Troy Empire.

In both computer software and ancient mythology, Trojans give attackers a backdoor to bypass the security and take over control. Back Orifice is a popular Trojan horse for Windows that does exactly that; it gives the hacker the ability to remotely take control of your computer.

There is also a new breed of **malware that leverages Artificial Intelligence** in its advanced functioning. This move completely changes the cyberwar landscape, as the attacks can be much more targeted and pass undetected for an extended period of time until the exact target is reached. For example, AI malware can attach itself to a modern video-conferencing application and analyse camera images in real time. Malware will display no destructive and suspicious behaviour until a specific target person appears in front of the camera. AI models trained for facial recognition will recognize the target person and activate the malicious code. This type of malware can be especially damaging to the reputations of political and public officials by unexpectedly revealing something embarrassing about them in front of a massive audience.

DDoS ATTACK

A Distributed Denial of Service attack is one of the most common and powerful cyberattacks. DDoS is achieved by flooding the target's network with large quantities of unwanted traffic and spam. To be successful in a DDoS attack, hackers will typically first need to assemble a botnet, which is an army of infected computers called *bots* or *zombies*, which can be activated by the hacker anytime. Most commonly, zombies are infected with dormant Trojan horses that can be activated upon request and take over the communication channel. Hackers will collect their army of malware-infected zombie bots for months or even years in advance, or if they need a gang of these zombies on short notice, they can purchase them via the Dark Web.

On the morning of this deadly zombie attack, the zombies will activate in unison and start flooding the target's network with traffic. When done correctly, this will overload the target network with unwanted traffic, rendering it unavailable for its intended users. As a consequence, the entire network has to be shut down in order to fix the issue.

This attack most often targets official government websites and massive institutions that heavily rely on the stability of their network. Banks are the perfect target for such an attack. Compromising a banking system so that it has to shut down for even an hour will bring immense financial and reputational damage. For this reason, institutions invest millions of dollars every year to continue enhancing their cybersecurity to ensure their network activities are closely monitored at all times.

SPYWARE

Spyware is a form of malware that allows hackers to spy on target computers and sniff out sensitive communication, passwords, credit card information, industrial secrets, and any other valuable data the offending party may be interested in acquiring.

A common example of spyware is a *keylogger*, which registers every key pressed on the user's keyboard upon activation. Naturally, this method is very effective in stealing sensitive information.

RANSOMEWARE

Ransomware is a type of cyber blackmail. Attackers will steal the victim's data or lock the user's computer, often leaving a single screen displayed on the computer with the attacker's Payment or Bitcoin account where the victim is blackmailed into sending the demanded sum in order to gain back control over his computer and data. The malware is often well-designed – such that it is still effective if the user restarts their computer as it doesn't allow the restart to remove the ransomware lock screen. At this point in the hostage situation, the malware could only be remotely disabled by either the hacker or professional anti-blocking software.

Ransomware is on a steady rise in the institutional world. It is common practice for cyber-attackers to deny a company the ability to operate effectively until either the requested sum is paid, or the company changes its policies. The latter is usually requested when attackers don't seek a monetary reward for their interference but are supporting a certain ideological change.

Ransomwares are also a very popular attack vector on local governments. Let's take a look at a case study on the way U.S. cities respond to these types of cyberattacks. In March 2019, Jackson County of Georgia paid a ransom of $400,000 to regain control over their systems. Garfield County of Utah was also subject to different cyber ransom the month prior.

While it may seem like a bad idea to give into the demands of a blackmailer, the reality is that in situations where the ransom is not paid, it usually amounts to higher losses. For example, in 2018 the city of Atlanta refused to pay a ransom of $51,000, which ended up costing them more than $9 Million in damages.

Looking for another juicy story on ransomware? Then look no longer than the case of *Eurofins*. Eurofins is a large European provider of pharma, food and environmental lab testing services, employing 45,000 people in over 800 laboratories. Here is where it gets tricky: Eurofins also offers cybersecurity assistance and is the UK's biggest provider of forensic services. Eurofins is used in over 70,000 investigations each year, which includes cases of cyber forensics. Despite being a cybersecurity service provider, their expertise in the field wasn't enough to save Eurofins from being a target of a catastrophic ransomware that brought down the majority of its operations. The official ransom amount paid to the hackers remains unknown, but it's safe to assume that Eurofins paid no small price to regain control.

PHISHING

An imposter e-mail, marketing campaign, or congratulatory message sent by hackers to trick their victims into sharing personal names, credit card numbers, birthdates and other sensitive information.

Chances are that at least once in your life you have received an email congratulating you for winning a million dollars in the national lottery, with the email correspondent kindly asking you to enter your bank account information so that a transfer can take place. That's phishing.

Machine Learning is being used increasingly by hackers to select the most effective campaigns. Meanwhile, anti-spam software increasingly relies on ML to better defend its customers. It is a cat and mouse game where the user is the coveted piece of cheese.

Here's an example of a phishing email:

"NATIONAL LOTTERY

Customer service

Ref: RNXF/34548121548451

CONGRATULATIONS!!!!

We are proud to announce to you the draw of the US NATIONAL LOTTO program held on 15th of April 2020. Your email address attached to ticket number 34548121548451 drew the lucky numbers: 15-4-22-10-12, which entitles you to the 2nd category of prizes. Congratulations! You have therefore been approved to claim a total sum of $760,000 (Seven Hundred and Sixty Thousand United States Dollars) in cash.

All participants have been selected randomly by the courtesy of the US Government from the World Wide Web site through a computer draw system extracted from over 100,000 companies and small businesses, to support the economy hit by the COVID-19 crisis.

The winning amount would be released to you by

our payment provider Worldlink. Worldlink agents will immediately commence the process to facilitate the release of your funds as soon as you provide them with your contact and payment details.

IMPORTANT: For security reasons you are requested to keep your personal and financial information safely and to not communicate it to anyone but Worldlink's agent who will contact you by phone as soon as you send them your contact details. This is also a supplementary precaution measure to avoid double claiming and unwanted abuse of the lottery program.

To file your claim, please contact Worldlink's fiduciary agent Mr. Paul Smith at richard.smith.worldlink@hotmail.com"

While it may seem silly to think that anyone would fall for a scheme like this, in reality thousands of innocent users fall victim to these kinds of phishing schemes every year.

FIREWALL

A firewall is one of the most important defence mechanisms standing between your private network and the open public network. Essentially firewalls contain rules on who is and who isn't authorized to access your network.

Once the attacker **bypasses** the firewall, it becomes useless and the only hope is that your anti-virus software will detect malicious code or activity going on.

There are three major ways to defeat a firewall:

1. Go around it – most companies have large networks with lots and lots of connections. Often not all connections get listed in a company's IT books and as consequence its firewall, some devices are forgotten or simply unnoticed, such as a connection of the newly arrived printer. If a hacker finds a connection which gives access to the network, their job is done.

2. Sneak through the firewall pretending

you are an authorized user. Depending on the firewall it is either easy, difficult or next to impossible. Badly configured firewalls leave vulnerabilities that can be exploited.

3. Take over the firewall. Again, it depends on the firewall software; it may be easy, or it may be very hard even for a super hacker.

As a little historical context, the first concept of a firewall was used on trains powered by coal. In the early days of coal-powered engines, sometimes the fire would break out beyond the engine-room and spread to passenger compartments. In order to prevent such disasters from happening, engineers literally started building an iron wall in the engine compartment to prevent fire spreading to other parts of the train.

"Social engineering bypasses all technologies, including firewalls."

- Kevin Mitnick, most wanted hacker of the 90s

ISO 27000 SERIES

ISO 27K (also known as the ISMS Family of Standards) is a series of best practices to help organizations improve their security measures. The advantage of every major Cloud provider over your own IT infrastructure is that they are all ISO 27k certified. If you're a business that uses IT extensively in day to day operations, obtaining ISO 27K certification or moving to the Cloud is an essential step toward ensuring a secure business environment.

CRASH PROJECT

CRASH, a new system of computer network defence that closely resembles the human immune system, is being developed by the US Department of Defence. CRASH is a promising attempt to massively improve the conventional firewall antivirus defence architecture. Inspired by the way the human immune system wards off infections, CRASH will allow computers to recover and repair themselves after a virus attack.

An issue with conventional antiviruses is that new malware is very good at mutating or shutting down once they detect an incoming virus scan. This behaviour is similar to a "free rider" on a train, who will run and hide in the bathroom as soon as the ticket inspector enters the compartment. With this new CRASH approach, every passenger on the train will be visible simultaneously upon the ticket inspector's arrival. Hence an intruder will not be able to remain unnoticed inside the system for very long.

BIOMETRICS

Biometrics is a way to securely authenticate into a computer system, but not by using a standard password or a key, which can be copied or discovered, instead the user is supposed to use his voice, face, fingerprint or any other type of personal biological data to login. Biometrics work well because our biological properties are unique to us. For example, there are no people in the world with identical fingerprints, allowing for a unique identification. IPhone started to include a fingerprint authentication option since IPhone 5S, which is a very handy option indeed.

NANOTECHNOLOGY

A branch of engineering working with technologies the size of a single bacteria. Nanotech is a slowly emerging science instituted with the ambition of creating a new line of products that touch on nearly every facet of life. Today nanotechnology is commonly used in electronics, most notably nano-scale transistors, which function as small switches that make computers. As time goes on and nanotechnology matures, we will increasingly see it overtaking currently advanced medical solutioning, DNA sequencers, nano-biological weapons, self-repairable material, ultra-resistant surface coating, nuclear fusion, molecular food printing and many other exciting aspects of the industry.

One of the most promising fields of nanotechnology is quantum computing. For the cryptology world, the emergence of quantum computing is a glitch of sorts, one where a discovery in physics can scrape decades of research in mathematics into the list of hypotheses proven wrong by science.

The next code breaker is coming from a place where cryptologists least expected in the form of a plethora of spinning electrons which crack the *uncrackable* codes in no time at all. Materials of atomic scale have been concealing more super computational power than you could ever imagine. Researchers from IBM, Google and NASA have finally managed to create quantum computers. It is no longer a matter of if, but only of when their powers will truly break the internet. That is, until new anti-quantum cryptology puts on its superhero cape and comes flying to the rescue.

"Nanotechnology is really interesting to me. Stuff to sort of make our world a better place, and a cleaner place, through science. And it also explains things that are happening. I've always been into it."

-Jessica Alba, Actress

QUBIT and QUANTUM COMPUTING

What is a Qubit? You guessed it – it's a Quantum Bit. A Qubit is a basic unit of information on a quantum computer. Just like bits on normal computers, qubits are used to store, transmit, and process information, but in much higher quantities.

In a nutshell, in order for this process to take place you must first lower your room temperature, and specifically a few atoms, to -273°C, which is near to absolute 0, the lowest temperature in the universe where everything freezes, atoms stop moving, and only electrons remain free. An interesting particularity of a qubit is that during calculations it isn't limited to taking values of 0 and 1, but also all the values in between thanks to its nanoscale which gives it these essential quantum properties. This factor significantly helps with calculations, thus making quantum computers vastly more efficient than classical computers.

However, contrary to classical computers which are always exact in their calculations, Quantum computers take a probabilistic approach to solving problems. Quantum computers simultaneously scan all possible results and output a statistical distribution of the possibilities. The most-probable results are then verified by the classical computer, which, as mentioned, is always exact in its calculations. This makes quantum computers ideal for breaking cryptological keys as well as for training Artificial Intelligence. In essence, these computers carry enough concerns to create the next technological revolution.

IBM and Google have already released their first functioning Quantum computers. Google even claims that their quantum computer has already achieved quantum supremacy on a family of new tasks. Quantum supremacy means getting results faster than currently available on the fastest classical supercomputers in existence. Their long-term goal is to offer Quantum services via their Cloud Platforms.

A technological revolution is already on the way and the age-old question remains: which companies will adapt and continue to grow, and which ones will refuse to implement the newest technology and be left in the metaphysical dust that Quantum computing leaves behind?

COMPUTER BASICS
FOR EVERYONE

PASCALINE – HISTORY

Pascaline was the first version of a modern automatic calculator, making it the oldest direct ancestor to a modern computer. Sometimes referred to as the "Arithmetic Machine", the device was named *Pascaline*, after its creator, Blaise Pascal. Pascal was an 18-year-old French mathematical prodigy who created the mechanical device to speed up his father's accounting work.

Many iterations of this concept would follow, all the way up to the little pocket-sized supercomputer you routinely use to swipe through Instagram, compose and read texts, and manage phone calls.

ENCRYPTION

Encryption is method of securing communication by altering an original message in such a way that it becomes impossible to read. The only way to read the message would be to reverse its encryption with a secret key. Encryption is the cornerstone of any modern communication system and is the fundamental building block of the internet.

"War is 90% information"

-Napoleon

As pointed out by the French Emperor and great military commander - Napoleon, having the right information at hand is fundamental in any warfare, hence diligently hiding important information and securing communication is crucial.

One of the oldest known examples of encryption is *Caesar's cipher*. This encryption technique consisted of replacing each letter in the message with a different letter according to a predefined key, in the absence of which it becomes virtually impossible to decrypt the message. This genius mathematical invention was used in ancient Rome, and as the name suggests, was implemented to encrypt top-secret messages between Caesar and his generals. As time has gone by, skilled cipher crackers have become skilled at cracking monoalphabetic-based encryptions, thus creating the need for more sophisticated methods.

In general, mathematicians in the field of cryptology spend years in their labs searching for new and ingenious ways to encode communication. Once they believe their encryption method to be strong enough, it is then passed on to professional encryption-crackers to see if they can break it. Some of the most widespread modern encryption solutions which have withstood time include public key cryptology, one-way hash functions and digital signature schemes. Everything we do on the internet is secure, thanks to those inventions.

The biggest employer of cryptologists in USA? – The NSA.

ENIGMA CODING MACHINE – HISTORY

The Enigma is an encryption machine developed in the early 20th century to protect commercial, diplomatic, and military communication. It was used extensively by Nazi Germany during World War II.

Enigma was typically operated by two people. The first and primary operator slowly types out the text in standard language. Every time a letter is pressed, it triggers a specially conceived set of rotors that do their magic by highlighting a different letter referred to as a ciphertext letter. The second operator will then record the ciphertext.

Military enigma had 158,962,555,217,826,360,000 different settings, which earned it the prestigious reputation of being *uncrackable*. This reputation persisted for a long period of time, that is until English computer science prodigy Alan Turing and his team of mathematicians at Bletchley Park designed an algorithm that would be operated by a machine specifically designed to intelligently try many different deciphering combinations until the right one was found. Luckily for Britain, this process worked. The famously uncrackable Enigma had finally met its match.

ALGORITHM

An **algorithm** is a set of instructions with an exact approach to deal with a given task. Algorithms are much like the music notes you must play in the right order to achieve the right tune. By working together, narrow algorithms can construct a more complex algorithm, orchestrated by a main algorithmic conductor.

A pumpkin pie recipe is a good example of a day-to-day algorithm. You follow step-by-step instructions, simply written in plain English, to complete the task. However, most algorithms are created for computer systems and are encoded using programming languages specifically designed for computers to interpret and execute.

As you might imagine, algorithms are the core tool for automatically completing a partial or full set of tasks. They are our most direct way of explaining to a machine exactly what we want it to do.

British scientist Alan Turing pioneered modern computer algorithm theory. In the early part of WWII, Turing designed an algorithm that could decrypt the supposedly uncrackable code created by the

German Enigma machine. The algorithm Turing designed for his Bombe machine gave the Britain a key advantage in WWII.

It is important for a computer algorithm language to be transportable, which means it must have the ability to be executed not only on a single machine but ideally on every computer. The most common words used to compose algorithms are: "Do until...", "If such and such... then do this…otherwise do this..." Easy as pumpkin pie.

AI is a self-programmable machine, meaning it designs its own algorithms to resolve tasks by simply observing what somebody else does.

SYSTEM

A system is a set of separate entities working together to form a whole. The purpose of such representation is to study the whole, both from the inside and the outside.

Systematic thinking is a hot topic in the engineering world. It allows engineers to deconstruct complex concepts into smaller parts in order to understand their functions. If the system is too complex from the inside, we can further divide the parts into subsystems to study each of them individually. Nearly everything in the universe can be thought of as a system:

- A car is a system made up of an engine, steering wheel, gears, and seatbelts
- Earth is a system of continents, oceans, winds, and living organisms
- A human is a system of organs, a brain, a central nervous system, bones, and muscles
- A government is a system of executive, legislative, and judiciary powers
- Cities are systems of local commerce, housing, and laws

- A computer program is a system of various algorithms executed in ordered sequence
- Computers are systems primarily made of CPU, memory storage, and a keyboard (everything else is an accessory)

Anything can be considered a system. We are systems who live inside a bigger system. In order to study complex systems, especially mechanical or economic systems, engineers and economists traditionally draw them out on paper, with almost every system starting as a big empty rectangular box. Then they draw inputs that go into the system from the outside world. Inputs are followed in their journey through the system until they come out of the system, usually in a modified form. The latter are called an output, and of course we draw them too. This is where we reach the most interesting part – we have now closely followed input from the moment it enters the system all the way through its process to the output. This is what allows us to study systems from the inside-out.

Systemic thinking is an important tool in innovation research. It allows us to mentally dismantle a product or a production process into many parts and see which of them can be improved. Sometimes engineers employ systemic thinking to fix problems. They may find that the expected output of the system is either deformed or didn't occur at all, meaning there is a block inside the system.

System analysis allows to understand a system in its entirety to then replicate it by encoding into a machine the algorithm by which the system works.

COMPLEXITY

Complexity is a measure of the diverse behaviours that a system can potentially display. The more complex the system, the more unpredictable it becomes. However, we find a hidden sense of order and harmony in most systems after diligently studying their behaviour.

Measuring and understanding the complexity of tasks is *crucial* in computer science because it provides the primary metric for how much computing power is required to resolve a task. As you can imagine, depending on the nature of the task, complexity can be exactly measured or subjectively qualified. For example, we can exactly quantify the complexity of a task required to sort a list of a million words in alphabetical order. However, the complexity of tasks such as creating a PowerPoint slide or writing a poem is much more subjective. Also, tasks considered easy for humans are not necessarily easy for computers and the other way around; some tasks require prior *experience* to be resolved.

We humans often possess decades of experience in task resolution whereas computers are just very fast calculators with no experience. AI, however, is shifting that paradigm, emerging as computers with *experience*. The more a task is reliant on experience the harder it is to judge its complexity in mathematical terms. We hope that future AI will be as good at solving experience-related tasks (from creating better PowerPoint slides to more pressing problems such as protein folding control to resolve cancer) as computational tasks. Indeed, besides the possibility of Terminator-like scenarios, AI can also contribute to a future without disease, poverty, or suffering – solving impressively complex problems along the way.

COMPUTER

A computer is a large calculator with attached memory storage. Computers are useful for their capacity to run advanced algorithms, the ability to listen for commands from various input channels, and their ability to send out commands or communication to other computers. World's computational capacity is doubling every two years and is opening possibilities for resolving ever more complex tasks.

Fun fact: The first electronic computer, ENIAC, weighed more than 25 metric tons and took up 167 square meters. Your smartphone is a trillion times faster, many times lighter, and definitely much smaller.

BUG

A computer bug refers to a badly coded function in computer software. To prevent future disruptions of the services or worse, it is essential that bugs are fixed as early as possible in the coding process. More bugs can accumulate in a snowball effect, leading to a total system crash.

Why a bug is called a bug? The fact is, our first computers were made out of big electrical circuits; the interface that allowed the user to see the results of what was happening inside their computer relay systems was a system of rows of connected light bulbs. Literally, just a screen made of light bulbs.

A light bulb that was turned off corresponded to a 0; and if was turned on, it corresponded to a 1. And it wasn't just five or six light bulbs tucked inside the computers, but hundreds upon hundreds. I'll give you a moment to imagine that. There was one problem, however. If one of the numerous light bulbs burned out or the relay system inside the circuit malfunctioned, then the entire experiment was broken and had to be restarted.

In 1947 an actual bug was caught inside a relay system, probably attracted to the light and heat. The bug caused the system to crash but has been successfully removed and the system was working again. The term bug was coined.

The biggest fear with relying on AI systems is that they are very hard to assess for bugs, allowing for scenarios like the one in Stanley Kubrick's iconic movie the *"2001: A Space Odyssey"*, released in 1968, just when the space race was in full swing. No more spoiler here, watch it over the weekend if you get a chance.

PUNCH CARD – HISTORY

The punch card was the first standard method of storing *tabular data* for automatic processing. It is an actual stiff paper card with 80 columns and 12 rows. To store information, holes are punched in specific areas. Think of it as a table containing 12 possible answers for 80 questions, such as marital status, age, income range, and so on. The idea to utilize punch cards in the computer science industry came from the textile industry's punch cards, which added massive automation to the weaving process decades earlier.

The need for automatic data processing came originally from the US government in 1890; assistance in counting and summarizing the census data of its rapidly growing population was desperately needed. The job was estimated to take 10 years to complete, which was not a viable solution due to the fact that it would overlap with the beginning of the next census count.

The eventual solution came from American inventor Herman Hollerith, who had just patented an electric tabulating machine that could automatically process punch cards and output statistics. Punch cards turned out to be a huge success and reduced the census counting task to a mere six years. Hollerith went on to establish the company we know today as IBM.

BIT

A computer **bit** is the most basic unit of information in all of information theory. There is no piece of information of a finer granularity than a single bit. Bit stands for binary digit, with binary indicating only two possible values in the numerical alphabet: 0 or 1.

"Bit by bit"

–wisdom quote

Not convinced that you can pass meaningful information using bits? Here is a Byzantine invention for communicating a single bit that helped the Empire last for centuries: A series of towers were built throughout the Empire's territory with only one aim, to light up their fire if an enemy is spotted. A single bit of information, encoding a simple message, fire on = enemy, fire off = no enemy. In case of an enemy coming, fires would light up one after the other, and information would very quickly travel from tower to tower all the way to the capital, allowing the king's generals plenty of time to prepare for defence. Genius.

If you increase the number of bits, you are essentially able to encode any message in the universe. This fact was theoretically proven in the early 1940s by Claude Shannon when he observed a relationship between Boolean algebra and telephone switching circuits. Bell Labs hired Shannon to devise the most efficient way to transfer information over wires. In 1948, Shannon penned *A Mathematical Theory of Communication*, essentially laying the foundation for the digital age. This feat earned Shannon the prestigious title, "Father of Modern Information Theory".

Invention of bits incredibly simplified communication methods, storage, and data manipulation, which is why bits became the de facto language of the *classical* computer system. And the reason *classical* computer systems are emphasized is due to the emergence of quantum computer systems. Entering the computer science arms race, quantum computer systems operate vastly differently from the *classical* systems – quantum systems use qubits instead of bits. But that's a story for another day.

Some maths on a classical bit: An array of 8 bits is a byte. 1024 bytes is a Kilobyte (**Kb**), and 1024 Kb is 1 Mb, and so on. Yes, in computer science bytes are multiples of 1024 instead of 1000. This is due to the fact that one bit stores two potential states – 0 and 1.

Two bits stores four potential states: 00, 01, 10, 11. Three bits stores eight states and so forth, so at the end you have an exponential increase in memory in the following way: 2, 4, 8, 16, 32, 64, 128, 256, 512, 1024.

You understand now why all those numbers appear more frequently in the computer industry instead of usual round numbers – it's because they are related to the amount of information that can be either stored or simultaneously handled by the processor.

Super computers of the past century generally had 1 to 100 kilobytes to deal with. Today, we produce as much as 10 exabytes of data a day. That's 11, 258, 999, 068, 426, 240 kilobytes.

ASCII

ASCII is the international standard for character encoding in computer systems using 8 bits. In other words, ASCII is a table which associates every keystroke on your keyboard to one of the 256 different numbers possible to encode using 8 bits.

For example, the number 3 is encoded as 0110011 and the letter "a" is encoded as 1000001, and this is the case on every standard computer. Every time you type a letter, a corresponding sequence of bits is sent to the processor all at once to inform it of the inputted character. Depending on the character, the computer will either move the cursor or type a letter or execute a function.

Here, encoded in ASCII, is Tolstoy's famous quote from *War and Peace,* "Nothing is so necessary for a young man as the company of intelligent women.":

ASCII: "01001110 01101111 01110100 01101000
01101001 01101110 01100111 00100000 01101001
01110011 00100000 01110011 01101111 00100000
01101110 01100101 01100011 01100101 01110011
01110011 01100001 01110010 01111001 00100000
01100110 01101111 01110010 00100000 01100001

00100000 01111001 01101111 01110101 01101110
01100111 00100000 01101101 01100001 01101110
00100000 01100001 01110011 00100000 01110100
01101000 01100101 00100000 01100011 01101111
01101101 01110000 01100001 01101110 01111001
00100000 01101111 01100110 00100000 01101001
01101110 01110100 01100101 01101100 01101100
01101001 01100111 01100101 01101110 01110100
00100000 01110111 01101111 01101101 01100101
01101110 00101110"

That's exactly how this quote is saved on your hard drive, exactly seventy-six times 8-bit sequences, including the last dot. When you save a text document, a mini laser creates a sequence of long and short magnetic bands on the hard drive, with long corresponding to a 1 and short to a 0. Then another laser can read it and display back on your screen when you browse to that location.

HARD DRIVE

A hard drive, sometimes called a hard disk, is an electromechanical device which allows massive amounts of digital data to be stored on magnetic discs. Today, the average personal computer hard drive stores up to multiple terabytes of data.

A computer's hard drive is the only reliably persistent memory storage system compared to other memory storages like registers and RAM, which are emptied as soon as they are powered off. In recent years, another type of memory has gradually started to replace hard drives – flash memory. However, it currently remains the more expensive option.

RAM

Besides being a major Hindu deity and the seventh avatar of the god Vishnu, RAM is also the second most important property of any computer, just after the CPU power.

Random Access Memory (RAM) refers to the short and midterm memory of a computer. RAM functions while the computer is working but is fully emptied out as soon as the computer is turned off. So why to waste time with a memory system that can potentially be lost in any power outage? The answer is speed; RAM is much faster than hard drives and is crucial for any computer system.

As soon as computer turns on its starts by uploading the most important data and programs into the RAM, to work from there rather than from the slow hard drives. If some of the lesser used data is needed, the computer will fetch it from the hard drive and upload to RAM while removing from RAM data it hasn't been using recently. RAM is very expensive hence its total storage is many times smaller than the hard drives, which is why we can never upload everything into RAM. In 2020 an average personal computer RAM is between 8 GB and 16 GB, while an average hard drive is between 512 GB and 1 Tb.

CPU and INTEL

The **Central Processing Unit** (CPU) is a powerful semiconductor chip that performs the role of the central orchestrator in any computer. Without a CPU there is simply no computer.

A CPU is sequential; it resolves one task at a time at a very fast pace and then immediately moves to the next task it has in the queue.

CPU power is doubling every two years, a phenomenon which is called Moore's Law, named after Intel's co-founder Gordon Moore. **Intel** is the most prominent designer and supplier of CPU chips in the world.

After many decades of CPU improvement, mainly consisting in ever smaller transistors, we are now arriving at physical limits of transistor size, hence the future of CPUs is being questioned. Intel is however looking into different ways to circumvent the limits. 3D shaped chips as well as quantum chips are amongst the most promising areas of research.

GPU and NVIDIA

A **Graphical Processing Unit** (GPU) is similar to a CPU but is designed to handle tasks which require much higher degree of parallelisation. Such tasks include graphics rendering and high-quality image display, heavy on matrix multiplications. As you may have noticed, AI training also relies on large matrix multiplications, making GPU a far better tool for the training process rather than traditional CPUs.

NVIDIA is the world's leading GPU designer and supplier. Since the rise of deep learning, NVIDIA stocks have skyrocketed, especially since the initial rumours of their involvement in embedded hardware for self-driving cars.

CPU and GPU work in tandem just like our brain has left and right hemispheres. One is more sequential and analytical while the other is more parallel and holistic.

SOFTWARE

Software is a collection of programmed algorithms that have been packaged together in a stable and easy to use information technology (IT) product. Good software is made of individual modules which have each been tested for stability, security, and ease of use.

Software is the direct opposite of hardware – hardware refers to the tangible components of a computer. Software is created using programming languages.

The most well-known software company in the world is Microsoft, notably renowned for their bestselling Windows operating system and Microsoft Office software suite, including Word, Excel, PowerPoint, Outlook, and more.

"That's the thing about people who think they hate computers. What they really hate is lousy programmers."

–Larry Niven, author of Ringworld

JAVA

Java is one of the most popular general-purpose programming languages on the planet. The basic version of Java is completely free for use.

Released by Sun Microsystems in 1995, Java went on to be acquired by Oracle in 2009 for a massive $7.4 billion, so that Oracle can offer an enterprise version of Java, sold for a fee. Java's motto has remained the same since its inception: "Write once, run anywhere," and thanks to its abstraction layer from underlying hardware, Java allows programmers to write a single piece of code that can be understood by any computer that has the Java Virtual Machine (JVM) installed. Many computer and mobile devices nowadays are sold with a JVM pre-installed. Android is fully built on Java. Anyone can write Java code and run it on his phone. For example, you can build a game or an application using Java. iPhone on the other side supports Swift language and not Java, but Swift is very similar to Java, just like Spanish is similar to Italian, so the concept remains the same. Outside the Apple world, almost no devices can understand Swift coding language, but they often do understand Java.

Despite its many functions, Java is rarely used to create Machine Learning applications, nor is it used in Data Science. Due to historical events, Python's community was more active in the Machine Learning field and has developed more tools that simplify AI engineering tasks. Nevertheless, Java is good at handling data pipelining and data infrastructure design. Hence Java is more frequently used by Application Developers and Data Engineers while Python is preferred by Data Scientists and AI Engineers.

DATABASE and SQL

A database is software which stores data in a *tabular format*. Databases are designed to be basic in their user interface but strong in their ability to operate at high speeds, contrary to personal table management applications such as Excel, whose priority is an easy-to-use interface. The original database was invented by IBM in their quest to fully digitize the punch card system.

Databases are crucial to the modern world, and if data is considered the blood of our economy, then the database is the crucial skeleton that holds all the data together. Every government and major business today heavily rely on databases to organize their most important data.

Common database applications include airline reservation systems, product inventory management, banking transactions, online purchases, and client information data. Most databases are kept on a single central server. Whoever needs to access the data has to submit a request to the system administrator for a username and password. With this information, the user can successfully retrieve the piece of the data required through the network without needing to download the entire database.

On a historical note, although IBM invented the database, it wasn't long before Larry Ellison, a computer scientist at the time, began working on a database project for the Central Intelligence Agency. After leaving the CIA, Ellison used his database expertise to establish Oracle, the leading database in the world for decades to come. By 2010, the Oracle database made Larry Ellison the sixth richest person on Earth and the third richest American, just behind tech-giant Bill Gates and famed investor Warren Buffet.

SQL stands for Structured Query Language, a simple and easy to use computer language invented by IBM so corporate administration employees could query databases without requiring the assistance of expert-level computer scientists. Today SQL is ubiquitous in the enterprise world and is the de facto language used to select data in a database.

BLOCKCHAIN and BITCOIN

Blockchain is a distributed database, which means every user has a complete copy of the database stored on their machine. It differs from the classical database which is managed by one central administrator and made accessible through the network to everyone with the valid credentials.

The point behind blockchain is that if everyone holds a copy of the entire database, then it would be vastly more difficult for any alteration of its content to go unnoticed. To successfully do this, one would need to hack every machine in existence and alter the data or do this for at least over 50% of the machines. This makes information counterfeiting much harder than it is on classical databases.

But the problem remains that blockchain is heavy on storage, due to everyone literally holding a duplicate. Critics have called this approach unscalable, not to mention that it is not the greenest, impact-wise. In the future, blockchain technology will be increasingly used to provide authenticity to physical or digital objects in order to prevent fraud and counterfeiting.

A well-known example of blockchain application is a crypto currency such as Bitcoin, which is used to conduct anonymous transfers of liquidity. Bitcoin was created in an attempt to fight back against the banking fractional reserve system, which had been accused of fabricating money and causing inflation. However, Bitcoin adoption is slow and still in a speculative phase due to the high volatility of its price.

"Bitcoin is better than currency"

–Bill Gates, Founder of Microsoft

FILE SYSTEM

A file system is the most common way to store digital documents. File types and data structures can vary freely and don't need to be all the same, contrary to the data stored in a database system. A file format in a file system can be of any format at all, Word, Excel, PDF, TEXT, JPEG, PNG, PYTHON, JAVA, etc.

No doubt you are familiar with the file system, as you use it every day whether you realize it or not. Whether you hop on your computer to search for old documents, family photos, or your favourite songs.

By the way, internet browsing works more or less the same way as your local browsing system. Essentially every website is a folder on some remote computer. What you are actually doing when you visit a website is that you are visiting the folder where the website is located. The file called index.html, known as the landing page, then gets automatically displayed in your browser.

At its surface the file system seems the same for Mac and Windows, however they differ greatly in the way they are saved in memory. I won't go into it here, but let's just say Windows is much more prone to viruses and data leaks contrary to Mac.

5G

The fifth-generation mobile communication technology that allows for unprecedented bandwidth in the likes of 10 GB/s, which is enough to stream Virtual Reality of quality indistinguishable from the real world.

5G is vastly more capable than its 4G ancestor, as it occupies a larger spectrum of frequencies allowing for more simultaneous communication channels.

Another peculiarity of 5G are its specially designed antennas which will supposedly follow users and focus energy emission towards their exact location rather than standard broadcasting equally distributed in the space. To achieve this goal, the antenna will first send a scanning array (like a sonar) to identify user location and then send targeted communication, making sure to not waste any unnecessary energy.

If you wonder if "you would possibly lose bandwidth when you move too fast" remember – data travels at the speed of light, which is a little less than 300,000 m/s, enough velocity to reach the moon in a mere 1.3 seconds. So naturally you will be easily tracked by the antenna.

IoT

"The internet will disappear. There will be so many IP addresses, so many devices, sensors, things that you are wearing, things that you are interacting with, that you won't even sense it. It will be part of your presence all the time. Imagine you walk into a room, and the room is dynamic. And with your permission, you are interacting with the things going on in the room."

–Eric Schmidt, CEO of Google

Internet of Things is a term which describes every possible device connected to the internet, from your pillow to your fridge. Even a lightbulb can now be connected to your phone via the internet, which makes for an easy and shock-free switch. Not speaking of your heating, which you can now turn on your way home.

IoTs are making increasingly large strides in our economy as essentially every service provider is trying to make their products "smarter".

The sudden increase in the volume of connected devices worldwide is certainly raising new risks with security. But fear not my friend with a smart-connected bathroom mirror. Companies are tossing out the big bucks to ensure their products do not make headlines for some notorious data leak, as their entire business reputation and market valuation is at stake.

AGE OF CLOUD

INTERNET

"I mean you can learn how to build a bullet or build a gun or build a bomb on the Internet."

–Barbara Bush, First Lady

The internet is the world's largest single system of interconnected networks, computers, and mobile devices – with the single purpose of sharing communication and computer processing resources over distances in a secure and reliable manner.

One hundred and sixty years ago, the first transatlantic telegram travelled from Britain to the United States along the first undersea wire. That telegram consisted of 21 words – and took seventeen hours to arrive. Although pulling such cable from one continent to another was an immense step forward and a true technological breakthrough, today, the same trip takes as little as 60 milliseconds while allowing for much bigger data bandwidth.

The modern internet was established, along with the World Wide Web, in 1990. The internet was initially invented so U.S. Department of Defense military units and a few top tier American research universities could share military and educational information. Remember that in those times personal computers didn't exist yet, and only corporations, governments, and the best educational institutions could afford to have a computer.

In the absence of tiny memory sticks, transporting gigantic memory storages to each other was a problem, especially when many other universities joined the network to share and access content. Literally, a wire connected two or more computers together allowing one computer to access the documents stored on the file system of another computer. Both computers would have to be simultaneously turned on, of course.

Millions of miles of thick fibre-optic cables are lying in intercontinental oceans. Belonging to governments, private companies, and cloud providers, those cables pump vast quantities of information across the planet. The McKinsey Global Institute reports that 543 terabits of data flew across borders every second in 2017. Interconnectedness of information – that is the internet.

HTTP and HTTPS

Hypertext Transfer Protocol (HTTP) is the standard communication protocol on the internet. HTTP defines the standard rules on how files can be sent in the most efficient and secure manner across the internet. You can think of it as a standard language that every resource connected to the internet understands.

HTTPS is a more modern version which also ensures that all data being communicated on the internet is encrypted, rendering it more secure.

BANDWIDTH

Bandwidth is the maximum amount of data that can be transferred or consumed over a given period of time. Previously, internet cables were made of copper and information travelled in the form of electricity, enabling up to 100kb/s bandwidth. Most internet providers have since switched to optic fibre, which transports information in the form of beams of light, enabling 1-10 Mb/s bandwidth on average. This is enough to download a high-quality movie or TV show in less ten minutes, a huge advantage over past eras when it took on average ten minutes to display a single image of Pamela Anderson on your screen, and you could literally see pixels slowly unrolling on the screen from up to down. Goodbye pixelated Pamela, hello Westworld.

It may feel a little challenging to imagine, but in reality there is nothing mysterious about bandwidth. Just like a gas pipeline, data bandwidth is limited to the size of the pipe. And just like a gas, data can be compressed; the more you compress it the more you can transport.

WEB

World Wide Web, or **www**, is a standard structure of information organization on the internet. Here's how it works: Each resource accessible on the web is given a unique www address, also known as an IP address, such as 191.255. 66.102. This address actually directs you to the country, then to the region, and finally to the actual physical address of the computer or server on which the desired resource you are seeking is stored. Original IP addresses are hard to remember, but thanks to the DNS system, important ones are translated into easy to remember domain names like "google.com."

The web system happened to be a delightful way to use the internet and effectively launched the e-commerce boom. Today, almost every transaction in the world goes through the internet one way or another. From ordering books on Amazon, to watching movies on Netflix, to swiping your credit card in Walmart, the internet is the consistent middleman.

"The Internet is the easiest thing to get into. To be an Internet retailer, you just get that URL."

–Bill Gates

BACKUP

Backups are widely used in police forces and SWAT teams, it is literally someone who has your back. Nowadays, however, we most often refer to **digital backups** – copies of files and data that are archived on a separate memory storage. Backing-up information is a way to mitigate the risk of damages, loss, or theft of the primary memory storage. Backups are crucial in the corporate and institutional worlds, where the loss of client data can lead to irreparable damage and imminent bankruptcy.

Just imagine if Bank of America's data center was hit by a tornado. What would happen to your bank account? Lost forever, and along with it your entire life savings? Fortunately, Bank of America backs up all its data and stores the backups in at least three different locations around the globe. All of the locations are protected with sub-military grade security.

The Holy Grail of memory storage would be the ability to back up our entire lives, which could be restored in the case of a fatal accident. For better or worse, with current computer-brain connection research projects, we could see this technology emerge within the next 30 years.

SERVER

A server is a computer which is upgraded with special software and hardware that gives it the ability to run for long periods of time without the need to be shut down and restarted like a normal personal computer. We are talking about months of work with no interruption.

Servers are the crucial 'building block' layers that make up the internet and allow it to be accessible without interruption. Most servers have backup twins, so in case one of them crashes, the twin server should still be able to serve its content to internet users.

DATA CENTER

Data centers are extremely large rooms, sometimes even entire hangars, filled to the brim with servers. Data centers often require big refrigerators to keep the place cool as servers produce a lot of heat. Most big corporations have their own data centers. Smaller corporations also often need data centers, but struggle to set them up.

Setting up a data center is an extremely complex engineering challenge. First, you need to find a region which is free of extreme weather conditions and preferably has a sufficiently cool climate. The region should also be politically stable, for obvious reasons. Once the place is selected, one needs to pull huge amounts of internet and electricity cables to the data center, supply all the computing hardware, storage hardware, and usually, refrigeration machines. Then comes the daunting task of connecting all the machines.

Once the hardware is plugged in, software engineers come in to set up software on all those machines, including operating systems, network software, drivers, firewalls, antivirus, and more. And finally, you need to physically secure the perimeter and set up a reliable maintenance structure in case something goes wrong – and when it comes to computers – something always goes wrong. In addition, for backup reasons that you already know, an almost identical copy of the same data center must be set up in a different region. Not speaking of backing up electricity supply with expensive local batteries and generators.

CLOUD

Cloud is a business model in which one can rent out a personal computer, memory storage, database, or an entire data center over the internet. With cloud, a guest user has the ability to log into a rented machine and fully operate it remotely over the web.

This is a very useful tool for both direct consumers and businesses. You and I, for instance, may enjoy having our files tucked away in OneDrive or iCloud with the ability to access them from any device that is connected to the internet. This is possible because our data can be stored in a remote data center belonging to Microsoft for OneDrive or Apple for iCloud, both of which are accessible by any mobile device, assuming we have the required login credentials.

However, it may be humbling to learn that retail consumers such as yourself are in reality, just a tiny drop in the sea of the cloud business. The most important cloud clients are businesses of every stature, ranging from very small startups to the largest corporations who simply didn't want or didn't have the capacity to set up their own data centers.

Cloud was pioneered by Amazon in the early

2000s, when it had already become the *Earth's Biggest Bookstore*. In order to serve all of its connected users, Amazon needed to acquire a staggering number of servers and data centers. However, Amazon noticed that customer interest in amazon.com was seasonally based. This meant that while sometimes all servers were running at full speed, there were other times when the majority of servers were sleeping. A living business machine, by the name of Jeff Bezos, decided that it could be profitable to rent out his servers during the periods when Amazon wasn't using them.

Because setting up a data center is a challenging task, and even more challenging to operate and maintain, many companies were more than happy to pay Amazon a fee to rent its data center while Amazon ensured the data center ran smoothly. This also allowed startups with strong ideas to scale up instantly with the help of Amazon's cloud. The simplicity transcended the business world. The process of renting 1 or 1,000 servers was achieved by a few clicks, making the system accessible to anyone with a single computer and credit card.

Cloud offerings grew quickly and are classified in three major groups: IaaS, PaaS and SaaS. Today, in 2020, cloud servers are a 300-billion-dollar business and are expected to keep growing annually at 30%.

Unsurprisingly, three major cloud providers account for the majority of cloud traffic: AWS by Amazon, GCP by Google, and Azure by Microsoft.

> *"If someone asks me what cloud computing is, I try not to get bogged down with definitions. I tell them that, simply put, cloud computing is a better way to run your business."*

–Marc Benioff, Founder, CEO, and Chairman of Salesforce

SOFTWARE as a SERVICE

Software as a Service (**SaaS**) is considered the most accessible out-of-the-box service, requiring the least setup and least technical knowledge to use. We all are familiar with SaaS services such as Gmail, Facebook, Spotify, YouTube, Netflix, Airbnb, UBER, or PayPal. All of those are Software as a Service programs, sold in the form of web pages or mobile applications.

SaaS is extremely simple to use, as there is no need to set up anything computer related, no renting a machine, no installing software, no need to upload any of your own software – you just sign up, and in some cases pay a fee, and the service provider will manage the rest. Most of those applications rent cloud resources from one of the three top cloud providers. In the case of PayPal, for example, they are hosted on Google Cloud Platform.

Many SaaS providers rely on machine learning in either one or many aspects of their operations. Easy examples are Facebook and YouTube. In order to keep running according to their content guidelines, they need to ensure that the content is appropriate. But as they literally have Terabytes of new incoming data at all times, no human workforce could potentially review all of it. This is where AI police algorithms are increasingly used to automatically scan and filter sensitive content.

PLATFORM as a SERVICE

Platform as a Service (**PaaS**) is one level below SaaS on a technical level. PaaS users are usually not final retail customers, but are typically companies wishing to run their SaaS via a cloud provider, like start-ups and application developers. The PaaS user rents out a fully operational machine with pre-installed standard software designed to satisfy the user's needs. An example would be a machine with 64 GB of RAM, 10 parallel Intel CPUs, an installed Windows OS, an antivirus, and a Python programming language software – fully operational in just two minutes.

With PaaS, a programmer can instantly run his applications in the cloud without worrying about Windows updates or physical hardware, as all of those details are managed by the cloud provider.

Perhaps most crucially, PaaS will also monitor the traffic to your application and ramp up additional servers with the exact same software automatically, if needed, so that users don't have to wait in a queue. This is a highly desirable feature of PaaS, as a queue will significantly slow down the loading speed and corrupt the user experience which will result in the worst possible outcome for a business: annoyed users.

PaaS was truly a game changer for cloud business. Those who managed to create an easy to use dashboard with plenty of PaaS services grew in size and status. Those who struggled to offer stable and user-friendly PaaS systems simply fell behind because their user numbers plummeted over time, such as the case with IBM Cloud, LoudCloud, SAP Cloud, and many others.

The number of PaaS offerings has grown rapidly in recent years which led to the emergence of subcategories, one of which is AI as a Service. This, of course, provides you with all the necessary tools to massively simplify the training of your machine learning models as well as integration of the trained models into your company's operations.

INFRASTRUCTURE as a SERVICE

Infrastructure as a Service (**IaaS**) was the first and most primitive contribution to cloud technology. Users basically rented empty servers that they could access through the web.

Although IaaS was a much better solution to circumvent the need of setting up your own data center, IasS still required users to install operating systems, antiviruses, programming languages, databases, and other custom software before the machine could become operational.

Users were also the only responsible party for ensuring that all of the required software was up to date. Because of these pitfalls, it became clear that many users were looking for a ready-to-use Windows or Linux machine. Luckily, with pre-installed programming language of choice, IaaS slowly led to the creation of PaaS which swiftly put companies without a good PaaS option out of business.

AWS, AZURE, GCP and ALIBABA

The three biggest cloud providers are Amazon (**AWS**), Microsoft (**AZURE**) and Google (**GCP**), respectively. AWS, launched in 2006 was the first Cloud business conglomerate and up until 2019 accounted for almost half of the entire market share, followed by Microsoft and Google. A new player form China – Alibaba Cloud is also slowly climbing the Cloud ladder.

As a funny fact, Amazon's CEO, Jeff Bezos was an early investor in Google and in 1998 provided them with as much as $250,000 in funds. Amazon was already up and running for four years then. Fast-forward to today, Google happened to be the fastest growing company in the history of capitalism and is now Amazon's most feared rival in the Cloud space.

Cloud is an extremely profitable business. In the case of Amazon, AWS cloud is notably its most profitable business sector, more than books. However, Amazon should think twice before getting too comfortable up on the cloud throne, as there is likely a CEO or two doing everything in their power to gain a piece of the market share.

In early 2020 studies, AWS lost some ground and now accounts for a little more than a third of the market share, while Microsoft, Google and Alibaba have gained some market share ground.

IBM is also planning to come back in the picture after their recent acquisition of Red Hat and appointment of a new CEO. IBM is also betting big on the Quantum computing as a Service that they plan to offer via Cloud.

SATOSHI SAN

AGE OF VIRTUAL REALITIES

USER INTERFACE

The term user interface (**UI**) mainly refers to the design and content of the web page that appears in the internet browser, as compared to the server-side code and databases which are only visible to the website owners and not the final users. The user only gets to see what he has access to.

"Digital technology, pervasively, is getting embedded in every place, everything, every person, every walk of life is being fundamentally shaped by digital technology — it is happening in our homes, our work, our places of entertainment. It's amazing to think of a world as a computer. I think that's the right metaphor for us as we go forward."

–Satya Nadella, CEO of Microsoft

The point of a user interface is to facilitate interaction between human and a machine. A good interface is extremely easy to navigate, while a bad interface just makes you want to tear off your hair and break everything around you while you are looking for a specific setting which you can't find. As the rule of thumb, an interface should contain no more than seven main elements not become overcrowded. This is because human short term memory can only hold seven objects at a time on average. Every machine, from a washer to a computer, has an interface that a user interacts with to set commands.

User interfaces, although historically mostly visual, don't have to be pictorial. In Google Home and Amazon Alexa, the UI is based on voice recognition and designed for the user to interact via speech. As voice-based UI matures in quality as well as security, we expect to use it as frequently as visual UIs.

The goal of the current UI research, of which Voice UI is a big part, is to tap a sleeping elderly population market. Few grandparents know how to use a computer to order on Amazon or to pay their bills electronically. Voice assistants might soon allow just that.

The main advantage to voice UI is speed. In a few seconds you can order a pizza by simply saying "Alexa, order one large Chicken Barbecue Domino's" rather than going to Domino's website where it takes a few minutes to click through your order. And don't worry, Alexa will ask you to specify if you want your pizza right away or later for dinner, along with any other pertinent information. Addresses and payment accounts are usually already linked to your Amazon account, to which Alexa has access.

JavaScript is the most widespread user interface coding language on the planet and is also known as the language of the web.

Wireframing is a process of drawing a mock-up of an application with all its user interfaces and then specifying different interactions that happen when the user triggers particular events within the UI, like clicking on buttons. Wireframing is important to do prior to coding as it can point to potential flaws in the UI as well as difficulties of use. Since it is much quicker to process through than coding, it can save coders massive amounts of time and bring clarity to the business vision.

A/B testing is essentially an experiment where two or more variations of a product or UI are shown to users in order to determine which version is better. In the case of e-commerce, better means higher propensity to buy, also known as the conversion rate.

"UI wireframing, product prototyping and A/B testing are invaluable tools that every aspiring entrepreneur must use to his advantage"

-Jad Fayad, Entrepreneur

VIRTUAL REALITY

"VR is a way to escape the real world into something more fantastic.
It has the potential to be the most social technology of all time."

\- Palmer Luckey, Founder of Oculus Rift

Virtual Reality (**VR**) is specially designed computer hardware and software allowing for a new generation of immersive user interfaces, aimed at erasing all differences between a true reality and a simulated reality. The Holy Grail of VR is for the experience to be as good as a lucid dream, indistinguishable from the natural habitat.

VR is poised to change the gaming industry, education, e-commerce, and communication. Although VR quality has massively improved in recent years, unfortunately most innovations were in the visual environment. The VR experience representing other human senses and perception is lagging behind. Until we can smell and feel, in addition to seeing and hearing, we won't be able to truly experience VR.

AUGMENTED REALITY

Augmented reality (**AR**) is specially designed computer hardware and software aimed at enhancing a user's experience with a superimposed computer interface. The most common AR example is 'smart glasses' which can recognize objects and display information on the transparent screen located inside the lenses. For example, a person can be walking around a museum wearing AR glasses and an artwork's description, with contextual information, can appear on the lenses. Or you could be a soldier on the battlefield and your AR glasses could indicate the specs and weakness of the enemy tank approaching you. AR is basically Iron Man helmet display stuff.

"There are a lot of issues that it opens up like for example who is allowed to place an augmented reality object where. Can someone from Burger King walk into a McDonald's and place a Burger King ad there? What are the rules? It's a whole new kind of field of consideration."

-Scott Belsky

Just like VR, AR is having difficulties penetrating the consumer market at this stage. But that can change at any time – as quickly as the appearance of the iPhone pushed Nokia out of business. Intel and Qualcomm are expected to go toe to toe in what appears to be a high-stakes battle over AR/VR and AI supremacy.While Intel is set to sponsor the 2024 Olympic Games and will provide exclusive VR apps to users worldwide, Qualcomm was going to be acquired by Broadcom for 117$Billions, but it was blocked by Donald Trump for security reasons.

HOLOGRAM

A **hologram** is a device capable of displaying 3D objects without the need of a VR or AR wearable. The hologram experience resembles 3D cinema technology except that you don't need to wear 3D glasses.

Illusion is possible thanks to the physical phenomena of diffraction, creating a three-dimensional light field that gives the projected image a sense of depth. Products like Microsoft Hololens, which combines AR and hologram technology, are currently available, but their lofty price is a big stopper for penetrating consumer and business markets.

ALLEGORY OF THE CAVE –
PHILOSOPHICAL STORY

Plato's **Allegory of the Cave**, *composed in 520 A.D., is* an educational and deeply philosophical little story that touches on how reality is perceived differently by each creature. In summary, Plato describes a group of prisoners who were captured and locked in a cave all their lives. The prisoners have never been outside the cave and are not even aware of its existence, since all of their lives they were forced to stare into the cave wall in front of them. Occasionally, shadows appeared on the wall along with sounds. The shadows were the visitors passing in front of the cave but never entering the prisoner's area. One day one of the prisoners managed to break out, but as soon as he stepped out of the cave, he was blinded by the shining sun. After some time his view adapted and he could see colors and even his own reflection in water. He was so shocked he couldn't believe what he saw. The prisoner ran back inside to tell the story about the weird *reality* happening outside the cave. Of course, no one believed him, called him crazy, and told him to go away with his foolish stories.

In this allegory, Plato pointed to the fact that we all

live like those prisoners, mistakenly believing that we know life, while the reality is much more complex and complicated. Plato's work was visionary, for centuries later we discovered that Earth is not the center of the universe, that there are unobservable creatures like bacteria, and that what we see is only the daylight but in reality light has a much broader spectrum allowing some animals to even see at night.

Thanks to the scientific community, hopefully today we are more humble about claiming that we know what reality is. After having our minds opened more than once, thanks to some new reality opening device such as VR or AR, we are becoming more accepting of the possibility that there is much more going on outside our caves than we know.

As AI keeps getting smarter, chances are it will help us to discover new and previously unimaginable frontiers of the Universe.

> *"All great Acts of Genius began with the same consideration:*
> *Do not be constrained by your present reality."*
>
> -Leonardo Da Vinci

THE MATRIX – MEDIA CULTURE

Scripted by the Wachowskis and released in 1999, *The Matrix* is an iconic fiction film depicting a dystopian future. Deeply philosophical and at the same time filled with action scenes, *The Matrix* tells the tale of the last human survivors of the Artificial Intelligence apocalypse and their struggle to save Zion, the last human city.

"We need guns. Lots of guns"

–Neo, The Matrix

The survivors are driven by a prophecy that Neo will save them. The problem is that Neo is imprisoned in the virtual world known as *The Matrix*, created by AI to sedate human minds while using their bodily resources.

The Matrix became an iconic film due to its provokingly plausible apocalyptic scenario. The movie *Terminator* is also iconically futuristic but falls short due to employing a time travelling device, which is physically impossible. *The Matrix* features nothing that goes against the laws of physics. Moreover, many hi-tech entrepreneurs think that the number of revolutionary technologies appearing in the movie are reachable within our lifetime. *The Matrix* has jokingly evolved from the fictional film genre to 'documentary'.

The film's title refers to the digital world simulated on the central server run by AI. The word matrix is borrowed from mathematics, where matrix is one of the most widely used objects to store and process numbers in a table-like manner. Matrixes are extensively used in mechanics, robotics, computer graphics, simulations, and machine learning training. In fact, a machine learning model is nothing more than a long series of interconnected matrixes which together simulate an artificial brain.

Thanks to *The Matrix* and *Terminator*, the wider public is becoming more aware of the potential dangers of AI. AI is spiking an increasing number of critical discussions that affect our planet. Some discussions have led to worldwide agreements on the abolition of AI in the weapons industry. These continuing discussions are important because AI funding will only be accelerating, and capitalists see great profit potential in AI-based automation.

NEURALINK

Neuralink is a company initiated by serial tech entrepreneur Elon Musk. Aim of **Neuralink** is to create a seamless interface between the human brain and a computer, just like in *The Matrix*. Neuralink technology is intended to massively empower human thinking abilities, helping us compete with increasingly smart artificial intelligence.

Although we are already using computers to empower our calculating and working capacities, Elon Musk has pointed out on the issue of the low bandwidth of communication between humans and machines. Typing on keyboards and clicking on a mouse are very slow processes – you can at best enter a few keystrokes per second.

In order to overcome the limitation and make the interaction quicker, Neuralink technology intends to connect computers through tiny electrical wires directly to the neurons inside our brains. In July 2019, Neuralink announced that the brain sewing machines and the electrodes are ready. The company now awaits FDA approval to allow human trials.

Such technology shall also allow us to back up our entire memories on computer storage, and further open gates toward the immortality agenda.

MARS 2022

SpaceX, space agency belonging to Elon Musk, plans to transport the first two cargo ships to the planet Mars in 2022. SpaceX also plans to send an extra four ships in 2024, two of which will be carrying a **human crew** this. The urgency to colonize Mars, as Elon Musk envisions, will alleviate some risks of extinguishing the total human population in case of an escalated nuclear war, asteroid hit, or an Artificial Intelligence apocalypse, or being wiped out by a pandemic.

Elon Musk has started Tesla and SpaceX after eBay has acquired PayPal for $1.5 Billion, which he co-founded. So if we do end up with a station on Mars by 2022, we can give a big round of applause to eBay and all American second hand buyers for indirectly sponsoring the trip.

"Becoming a multi planet species is way better than remaining a single planet species."

-Elon Musk

AGE OF DIGITAL ECONOMIES

BUSINESS PROCESSES DIGITALISED

Business processes, including production processes, financial processes, supply chain processes and innovation processes are the step by step algorithms for running a business. In other words, a repetitive set of tasks that needs to be done on a regular or seasonal basis to keep a business working. For example, procuring raw materials needed for production on every Monday morning is a business process. The procedure of turning on a nuclear power plant every evening as soon as solar panels are getting off the grid, is a business process. The procedure for boarding passengers onto an airplane is a business process. Elaboration of quarterly financial reports is a business process.

Once established, business processes usually don't change much for years. Many even become embedded in company's traditions. However, in the past two decades an increasing number of business processes have become digitized or automated. This is driven by constant attempts for further cost reductions, margin improvements and a search for better customer service.

Everything started with Toyota coming up with a

process called "Continuous Improvement". Which is basically a process of improving other processes. After years it all came down to either reducing waste, digitizing work, automating work or outsourcing work to a cheaper country.

"Resources are what he uses to do it, processes are how he does it, and priorities are why he does it."

— Clayton M. Christensen. Author of the Innovator's Dilemma

AUTOMATION

By making machines do all the work, **automation** reduces business and operating processes to minimal or zero human assistance. We first saw automation on a global scale during the first and second industrial revolutions; the invention of the steam engine led the way to partial automation of many heavy lifting and transport tasks.

In the past twenty years, we have seen a massive shift from paper-based processes to fully digitalized processes. The switch from paper to digital prepared the ground for a new automation revolution, called the fourth industrial revolution, which will shift society from the end of the industrial age into the new intelligent machine age. This type of automation will, for the first time, affect white collar labour, such as corporate and banking operations.

The transportation industry is the next candidate in line for a massive automation wave, potentially resulting in millions of job losses.

However, we humans proved historically resilient in dealing well with automation and finding new jobs. Automation is an old idea. In the second book of

Physics, Aristotle would introduce the modern concept of automation, referring to 'spontaneity' in nature, an event itself triggering its occurrence.

In another book, on Politics, Aristotle reasons on automation from a social angle:

> *"If every instrument could accomplish its own work,*
> *obeying or anticipating the will of others… and if,*
> *in a like manner, the shuttle would weave and the*
> *plectrum touch the lyre without a hand to guide them,*
> *chief workmen would not want servants,*
> *nor would masters need slaves."*

-Aristotle, book on Politics 300 BC

Using music as an illustration he pointed out that it was terrible to enslave people to make music, which in those times was undesirable and labor intensive, but as society needs music, someone must be enslaved. Truly remarkable ideas for a society living 2,000 years ago.

ROBOT PROCESS AUTOMATION

Robot process automation (**RPA**) is software capable of automating work done by an employee on a computer. There are different flavours of RPA software, some smarter than others. The best RPA software allows a user to record tasks on a screen that the software will encode. While a human completes a daily repetitive task, such as opening an e-mail and copying different fields from the email into some customer management software, RPA records all the tasks the human has done. Now RPA is capable of taking over control of the computer and completing the tasks on its own – clicking, typing, moving, opening, closing, saving, copying, pasting, calculating, and performing any other function it needs in order to accomplish the given task. RPA gained in popularity when banks and other process-heavy companies realized they could automate 50% of operational jobs at a fraction of the cost.

Your job is to send reminders? Download receipts and enter them manually into the company's official financial records? Or complete an online form based on a printed copy of a form? Slowly start reorienting yourself because these jobs can now be easily automated.

SATOSHI SAN

API

Does the term **Application programming interface (API)** speak to you? It's not surprising, as the name doesn't really speak to anyone, even advanced developers. Nevertheless, this technology is the next big thing in corporate operations automation.

In essence, API is a common standard of communication between two different machines and software applications. For instance, between you ordering a pizza through your Amazon Alexa and the delivery of your pizza to your doorstep, a long chain of applications talk to each other through APIs. Alexa's API sends your command to Domino's customer API which then processes it and sends the command to a connected oven. Once the pizza is cooked, the connected oven will send a notification to an automated delivery truck which will then deliver the pizza to your door and send a notification back to Alexa's API. Alexa will kindly tell you your pizza is ready. No person is needed. Most often, API communication goes through the internet and information is passed via HTTPS requests.

166

DEVELOPER

A **developer**, also known as coder or programmer is a person who writes computer software code. There are many different types of developers – some focus on a very specific coding language, like Java or C#, and others prefer to get to know a bit of everything. Developers are also responsible for an application's security.

Developers who code both, the User Interface and the application server logic are called Full Stack developers.

In the past two decades we see Fortune 500 companies and major Banks reducing their operation staff and increasingly hiring software developers. JP Morgan Chase bank and Goldman Sachs go as far as referring to themselves as technology companies rather than banking companies. JP Morgan Chase for example employs more than 40,000 developers and has an annual Technology budget of $11 Billion. They have also recently built the biggest world repository for Data.

DATA SCIENTIST

"It is a capital mistake to theorize before one has data. Insensibly one begins to twist facts to suit theories, instead of theories to suit facts."

— Sir Arthur Conan Doyle, Sherlock Holmes

A data scientist is a modern-day corporate detective/journalist armed with Python, Excel, SQL, and a big fat salary cheque. Besides being a data-driven journalist, a Data Scientist is also a data-driven consultant. This means that in addition to reporting his findings in a few visual charts, Data Scientist is also supposed to leverage the conclusions of his research to advise businesses on tangible ways to make their operations better. Better usually means more productive manufacturing, better personalised products, more effective marketing and all of that without spending a dollar on new equipment or staff.

Most data scientists start as data analysts, freshly graduated from statistics, engineering, or computer science school. Data analysts often begin their careers as mathematically inclined introverts, but as they continue to progress and improve their presentation and language skills, they become eligible for data scientist promotion.

Data Scientists are increasingly using machine learning, letting algorithms crunch all the data while they sit back and relax. Data scientists who are more passionate about architecting AI rather than browsing thought data become **AI engineers** and evolve toward Chief Technology Officer (CTO) positions, while others become senior data executives and evolve toward Chief Data Officer (CDO) positions.

"Data Scientist: The Sexiest Job of the 21st Century"

–Harvard Business Review

DATA ENGINEER

Data engineers are computer engineers who ensure a reliable computer hardware and software infrastructure so that **Big Data** can be successfully collected, stored and processed. Data engineers ensure that data scientists can get easy access to all company data, within their permission rights of course. We are often speaking about terabytes of data.

Today, having the right data infrastructure in place is crucial for public and private institutions. As data grows, infrastructure has to mature and adapt. Unfortunately, not everyone manages. For example, the FBI was very late on its data infrastructure upgrade and in 2001 still had the majority of its processes based on paperwork. A commission was set to investigate how the 9/11 disaster could happen. But they couldn't simply receive the right information from the FBI – it was extremely disorganized. At the conclusion, the commission reported "The FBI lacked the ability to know what it knew." A Senator told the Washington Post: "We had information that could have stopped 9/11. It was sitting there and was not acted upon.… I haven't seen them correct the problems.… We might be in the 22nd century before we get 21st-century technology."

170

When the Bureau was asked a number of uncomfortable questions, the FBI answered with "Nothing to worry about, we have a modernization plan already in the works," referring to the new Virtual Case File (VCF) system they started to design earlier in 2001.

VCF was an attempt to go from a paper-based to an electronic-based management system, but it was abandoned in 2005 after costing $170 million of taxpayers' money. Apparently, the FBI didn't manage to get the requirements well defined. The same year, the FBI hired Lockheed Martin to launch a new program called Sentinel. They were convinced it would work this time, putting a mere $451 million price tag on it. Lockheed Martin promised that Sentinel xwould be fully operational by 2009. On March 3, 2010, the FBI abandoned the project. And it is only after hiring the "Agile" IT project management guru Jeff Johnson that the FBI was finally able to complete the project. According to The Wall Street Journal: "FBI Goes Digital, After Delays." In August 2012, the FBI's Sentinel system was finally ready for use, replacing manual processes and older electronic management tools. Since then the effect of the Sentinel Information system on the FBI has been dramatic. The ability to communicate and share information has fundamentally changed what the Bureau is capable of.

Today new Data modernization plans are taking place all over the corporate world, with the majority of architectural work conducted by Data Engineers. Data Engineers usually have a lot of experience under their IT belts. Data Engineers arm companies with technologies that allow them to store and process Big Data in-house. On one side Big Data is becoming available for companies' Data Scientists to analyse, and on the other side Big Data is becoming available for the AI engineers to train smarter ML models. All of this thanks to Data Engineers.

AI ENGINEER

AI engineer is the most recent addition to the big Data trio, gaining increasing popularity over the traditional Data Scientists due to the fact that most companies are now entering AI industrialisation phase. Corporate IT infrastructure is becoming mature enough to accommodate Machine Learning models in its daily operations. This journey of catching up to Google and Facebook that many companies have initiated a few years back is finally starting to pay off.

AI engineers usually come from the Data Science background but during their career focus on expanding their broader computer science skills, contrary to the most Data Scientists who remain statisticians with laptops.

AI engineers focus on two major tasks – developing/finding the most powerful Deep Learning model possible and working closely with Data Engineers to implement those models into company's operations.

CONSULTANT

A consultant is a professional who provides expert advice to businesses, updating them on the latest trends in the areas of automation, client experience, advanced technologies, tax optimization, law, cyber security, and marketing. There are three main branches of consulting making up for the majority of consulting business: Strategy consulting, Technology consulting and Finance consulting.

Technology consulting has been the fastest growing consulting sector, with top firms like Accenture now employing half a million consultants. Other major advisory firms, like the 'Big 4' auditing and accounting firms, are energetically expanding their consulting branches. The most prestigious space for consulting remains strategy consulting, with top firms like McKinsey and BCG advising directly on the C-suite level. Contrary to investment bankers who advise firms on lucrative merger and acquisition strategies, strategy consulting firms focus more on dealing with competition the traditional way. They also advise on opportunities for expansion toward new untapped markets.

To hire a consulting firm, most often corporations

such as retail banks or public services issue a request for a proposal document, in which they describe the problem they are trying to solve or their request for advice. Consulting firms reply with a proposal, specifying their approach including CVs of consultants who will work on the solution, the price, and the estimated time to completion. The selected consulting firm will gain the rights to the project and will soon send its tech army to the client's premises to deal with the issue.

FREELANCER

A freelancer is a one-person working resource. A freelancer is either a professional struggling to become a consultant, or inversely, an expert who prefers to be self-employed rather than working for a firm. For example, if Obama came to Harvard to give a law class lecture for $100,000, he would probably qualify as a freelancer. Another common type of freelancer is a software developer who chooses to relocate to a country with cheaper cost of living and better weather, like the Philippines, and remotely works for an American or European firm.

ENTREPRENEUR

Popularized as new age cowboys, **entrepreneurs** explore industrial frontiers armed with their charisma and futuristic vision. In practice, an entrepreneur is a person who sets up a new business from scratch, most notably leveraging innovations and technologies borrowed from one field and applied in a different field. Entrepreneurs rely on venture capitalists for funding and all hope to one day create the next unicorn.

STARTUP

A start-up is a company created by one or several entrepreneurs which is in its early stages of operations. Start-ups usually aim at disrupting old fashioned businesses by leveraging new technologies and discoveries, which traditional companies are too slow to adopt.

Start-ups usually have a "built to scale" vision from day one, and often require substantial funding before they can transform their prototype into a viable product ready for the market. Every start-up wants to either become a Unicorn or be acquired by a big company.

A **unicorn** is a start-up that has reached a value of over $1 billion, such as Uber, Epic Games, WhatsApp, Airbnb, SpaceX, and more. The term was coined by venture capitalist, Aileen Lee in 2013, in reference to a mythical creature which is virtually impossible to find.

VENTURE CAPITAL

Venture capital is a type of financing that investors provide to start-up companies and small businesses with high potential for growth and good projected future cash flow. All venture capitalists hope they are investing in the next Unicorn.

IPO

*"If one percent of Aramco is offered to the market,
just one percent,
it will be the biggest IPO on earth."*

- Deputy Crown Prince of Saudi Arabia, Mohammed
bin Salman, in 2016

IPO stands for Initial Public Offering and is basically
the process of introducing company's shares onto a
stock market, making them available to everyone and
not only the private investors. IPO is the last stage of
growth for any company and takes place for many reasons. An IPO allows a company to raise a lot of cash,
often needed for an expansion and globalization of its
operations. It is also a great way for the early venture
investors to finally cash out and take their profits.

IPOs are conducted by major Investment Banks
such as Goldman Sachs and JP Morgan and usually
get to keep around 3% of the whole deal, which is a
lot. Launching an IPO is like launching new money.
Only exception is that this money is not guaranteed
by the government, in exchange, investor is rewarded
with dividends and hopefully growing capital valuation.

The biggest sector for IPO business in 2019 was Technology sector, with Uber, Lyft and Alibaba leading the way.

The biggest single IPO in history however is not attributed to a Tech giant but to an oil & gas company. Saudi Aramco's long-awaited IPO, which took place in late 2019, valued the company at $1.7 Trillion.

OUTSOURCING

Outsourcing is the migration of a working unit from local headquarters to a country with a cheaper labour force, such as India for software development; China for manufacturing; or Poland for administrative tasks.

One may think of outsourcing in reference to Adam Smith's theory of economic improvement through labour division applied on a global level.

Outsourcing became an integral part of business strategies throughout the 1990s. But as the labour providing countries went through economic booms, prices of workers have gradually increased and now for the first time are no longer economically viable for some home businesses. This is especially true for software development, where we are now witnessing a reverse trend where jobs are being insourced back to European and American headquarters. Increasing cost reduction thanks to automation and AI is accelerating the insourcing trend.

THE GIG ECONOMY

A **gig economy** is an economy which is increasingly reliant on flexible and temporary jobs performed by freelancers and consultants rather than traditional full-time employees. The gig economy phenomenon is growing worldwide due to the rapid evolution of technological and business landscapes to which many internal company employees have neither time nor wish to adapt.

As a consequence, companies prefer to hire temporary workers that they can easily get onboard for conducting digital transformation and automation projects then easily dispose of when budgets tighten, like during the financial crisis of 2008 or COVID-19 lockdown. Another advantage is that consultants and freelancers usually have a broader range of experience and are knowledgeable about the competition, which is valuable information that they can bring to the hiring company.

GAFAM

GAFAM is an acronym describing the big five giants of the web: Google, Amazon, Facebook, Apple, and Microsoft. GAFAM is used to describe a quasi-total dominance of the entire internet that these five companies have amassed, legitimately classifying them as digital oligopolies. There is an ongoing concern about their world domination and influence, especially due to the staggering amounts of data GAFAM collects and holds, including tons of personal data.

Even more concerning is their active involvement in the AI industry, as data is the primary and necessary source for training machine learning models. GAFAM is well-positioned to maintain a monopoly on the growing AI market, making it almost impossible for outsiders to break in. Google, Amazon, and Microsoft are also the biggest cloud providers, possessing quasi-unlimited computing resources needed to train the biggest AI models on the planet.

BATX is the Chinese version of GAFAM, composed of Baidu, Alibaba, Tencent, and Xiaomi. BATX was developed during the Golden Shield initiative, which was propelled by the Chinese government to deny GAFAM access to the Chinese media market.

DIGITAL TRANSFORMATION

Digital transformation describes a wave of business process transformations from the old paper and physical-based processes to the new partially or fully digitized processes. Banks, insurers, and public services are the biggest clients of the digital transformation consulting industry. After a company has gone through a wave of digital transformations, it is then ready to embrace AI and RPA-based technologies to further automate or augment the company's operations.

Digital transformations vastly modify customer experience, often for the better. The Holy Grail is, of course, a real-time service to the customer, like Netflix, which allows you to create an account and start using the service within minutes. Public services are also undergoing digital transformation and adopting the real-time service mentality. Soon, if you get flashed while driving too fast, instead of waiting a month for a letter from police, you will receive an instant mobile notification with an exact settlement fee. Filing taxes will also be much easier.

SELF-DRIVING CARS

A self-driving car is a vehicle capable of autonomously navigating from point A to point B safely and in compliance with road traffic regulations. These cars are usually equipped with cameras, proximity sensors, and an AI computer brain infused with millions of miles of driving training, both in physical and simulated environments. Just like AI learned to beat the world chess champion in past decades, today AI is capable of navigating a car better than most human drivers.

This field also includes self-driving trucks. For better or worse, this technology could put an end to the largest employment sector in the United States – truck and taxi driving – effectively rendering tens of millions of employees jobless. It is currently up to policymakers to manage the transition, but drivers should seriously be considering their exit routes.

ALGORITHMIC TRADING

Algorithmic trading is an automated trading process which does not require human intervention in any step of its operations. Trading algorithms use either quantitative equations or machine learning models to make sound trading decisions. Data used by trading algorithms often includes quarterly revenues, financial ratios, and key indicators of economic states, such as unemployment rates, inflation and GDP.

Increasingly, news crunching and media **sentiment analysis** are being used in effective hype-driven trading strategies.

Today algorithmic trading dominates over 80% of total daily stock market. Due to such large volumes, many restriction mechanisms have been imposed by regulators and new ones appear every year, aiming to avoid market disasters such as the flash crash of 2010, when algorithms organised a $ 1trillion dollar market selloff in less than an hour.

PRECISION MEDICINE

Precision medicine refers to a highly personalised medical treatment made possible with Artificial Intelligence. AI for precision medicine is unanimously considered the field which will have the biggest impact on the Healthcare industry in the next five years.

Numerous studies have proven ML's ability to match or exceed the diagnostic accuracy of human medical professionals, demonstrating its promise to avoid medical errors, which is a costly and frequent phenomenon in hospitals worldwide.

For example, diagnosing breast cancer at stage 0, as opposed to stage I/II results in $21,000 reduction in medical costs. DNA analysis, already costing less than $200 per person will allow to flag patients with higher risk of developing cancer at an early stage.

Medical AI tech will also have a direct impact on hospitals high labor costs: AI could automate around 33% of health practitioner tasks and 40% of tasks completed by health support occupations by 2030.

By 2025 AI could save Healthcare industry up to $600 Billion in costs, while providing faster and more accurate diagnosis and boosting patient satisfaction.

EXPONENTIAL GROWTH

Exponential growth refers to a growth process in which quantities increase in an accelerating way, like an explosion, contrary to linear growth which is based on a constant increment over time.

There are plenty of examples of exponential growths in nature and human lives, from bomb explosions to the growth of biological organisms. In the latter, growth starts with a single cell, which divides in two, then each of the two cells further divides in two and so on. As a result you start from a process where a single cell is produced in one minute, then a few days later you have thousands of cells being produced every minute and few days after that you have billions of cells being produced every minute.

In a linear growth scenario, the growth rate doesn't accelerate. A manufacturing factory is an example of linear production. Same maximal quantity of products will be added every day of the week.

Human population also grows at an exponential rate, and so do viruses during pandemics.

Another subject growing at an exponential rate is

technology. Since the first electronic punch card processing machines from 1890's up until present date, technology has been growing at an exponential rate with no breaks. Not a war, recession or a natural disaster did slow down the technological growth.

Co-founder of Intel, Gordon Moore, was the first to notice that computing power is doubling every two years, has been doing so for decades and seemingly will keep growing at this pace in the future.

Exponential technology growth is putting us on the straight line to technological singularity, projected to occur sometime in the next few decades.

TECHNOLOGY

We understand the term technology on an intuitive level but it is a little more delicate to put into words. Broadly speaking, **technology** is a set of tools, skills, and methods intended to augment human abilities or facilitate our work through partial or full automation of tasks. Technology was always a crucial aspect of human civilization, perhaps one without which civilization would not have been possible in the first place.

Weapons helped us fight enemies and establish total control over surrounding territory, agricultural technology helped us produce enough food to store and feed a growing population, medical technology helped us fight disease, space technology helped us fly to the moon, and communication technology helped us effectively store and share knowledge, pushing the boundaries of our intelligence like never before.

Technology is without a doubt one of the most crucial aspects of human development, even more than the effective organization is. Ants, for example, are impressively well organized, but due to the lack of technological development, their societies are stuck in progress. Populations that master technology and organization govern the world and those who lack them either fall into chaos or lag behind the development curve. It seems technology has formed a strong historical symbiosis with the human animal and increasingly infuses itself in every aspect of human life.

"I am telling you, the world's first trillionaires are going to come from somebody who masters AI and all its derivatives, and applies it in ways we never thought of."
–Mark Cuban, Billionaire

SINGULARITY

Technological **singularity** is a point in time beyond which history breaks from a seemingly linear improvement cycle and enters an uncontrollable and impossible to follow intelligence explosion phase, beyond which the future is by definition unpredictable. Singularity may be achieved when artificial general intelligence will learn to improve itself better than humans can improve it, with each new generation being extensively better than the previous.

Due to the digital nature of AI, its evolution can be speeded up to the point where it passes from an average human level of intelligence to a genius level of intelligence to millions of times smarter than Einstein level of intelligence in a short period of time. What will happen to humans at that point? No one knows.

The term singularity was popularized by the director of engineering at Google, Ray Kurzweil, in his book, *Singularity Is Near*.

The road to technological singularity began exactly when we humans began to qualify as humans, differentiating ourselves from other apes with a more powerful prefrontal cortex that gives us the ability for critical thinking. Most of the deep thinkers seem to conclude that technology is the road to survival, domination, and a better life. Reviewing the history of the undeniable acceleration of technological development, demonstrated by the likes of Moore's Law, singularity is judged to be a very plausible scenario within the next 30 years.

"One cubic inch of nanotube circuitry, once fully developed, would be up to one hundred million times more powerful than the human brain."

–Ray Kurzweil, The Singularity is Near: When Humans Transcend Biology

UNIVERSAL BASIC INCOME

Universal Basic Income (**UBI**) is a government's guarantee that each citizen receives a minimum income. Discussion on whether UBI will actually come into force is an important ongoing topic, especially as potential job losses and unemployment rates spike due to automation.

The concept behind UBI is to provide a minimum wage to every citizen on a monthly basis, no matter the citizen's employment status, profession, or personal wealth.

ABUNDANCE

Abundance is a theoretical situation in the economy where water, food, and goods are produced in great abundance with minimal human labour required. As a result, the essential items are made available to all at a cheap price, or even freely.

In essence, abundance is a hypothetical situation where one doesn't need to work to survive but works only out of a wish for personal expression. Peter Diamandis, a multimillionaire entrepreneur, author, and friend of Ray Kurzweil and Elon Musk, suggests the concept of abundance is soon achievable.

"Technology is a resource-liberating mechanism. It can make the once scarce the now abundant."

–Peter H. Diamandis, Abundance: The Future is Better Than You Think

REFERENCES

On the Origin of Species
 by Charles Darwin (1859)

The Expression of the Emotions in Man and Animals
 by Charles Darwin (1872)

Psycho-Cybernetics
by Maxwell Maltz (1989)

Simulacres et simulation
by Jean Baudrillard (1985)

The Republic
by Plato (380 BC)

L'an de grâce
by Pascal (1654)

Brave New World

by Aldous Huxley

1984
by George Orwell

Animal Farm
by GEORGE ORWELL

Life 3.0
by Rob Shapiro and Max Tegmark

How to Create a Mind: The Secret of Human Thought Revealed
by Ray Kurzweil

The Singularity Is Near
by Ray Kurzweil (2010)

Transcend: Nine Steps to Living Well Forever
by Ray Kurzweil , Terry Grossman (2009)

The Age of Spiritual Machines: When Computers Exceed Human Intelligence
by Ray Kurzweil (2000)

AI Superpowers: China, Silicon Valley, and the New World

Order
by Kai-Fu Lee

Secrets & Lies:Digital Security in a Networked World,
by Bruce Schneier

Computer Viruses for Dummies
by Peter H. Gregory

Applied Artificial Intelligence: A Handbook For Business
Leaders
by Mariya Yao , Adelyn Zhou , Marlene Jia (2018)

Artificial Intelligence: 101 Things You Must Know Today
About Our Future
by Lasse Rouhiainen

Artificial Intelligence and Machine Learning for Business: A
No-Nonsense Guide to Data Driven Technologies
by Steven Finlay

The Future of Work: Robots, AI, and Automation
by Darrell M. West

Homo Deus: A Brief History of Tomorrow
by Yuval Noah Harari

GDP: A Brief, but Affectionate History
by Diane Coyle (Princeton, 2014)

Bill Gates et la saga de Microsoft
by Daniel Ichbiah (1995)

Google story
by David-A Vise (2006)

Steve Jobs Poche
by Walter Isaacson (2012)

Elon Musk
by Ashlee Vance

Scrum: The Art of Doing Twice the Work in Half the Time
by Jeff Sutherland

Introduction to Artificial Intelligence
by Philip C. Jackson Jr. (1985)

Reinforcement Learning: An Introduction (Adaptive Compu-
tation and Machine Learning)
by Richard S. Sutton, Andrew G. Barto

Python Machine Learning
by Sebastian Raschka (2015)

Python Data Science Handbook
by J. Vanderplas (2018. O'Reilly Media)

Python for Data Analysis
by W. McKinney (2018. O'Reilly Media)

Machine Learning Applications Using Python
by P. Mathur (2019. Apress)

Deep Learning with Applications Using Python
by N.K. Manaswi (2018. Apress)

Foundations of Machine Learning
de Mehryar Mohri et Afshin Rostamizadeh

Linear Algebra
by Georgi E. Shilov (1977)

How Life Imitates Chess
by G. K. Kasparov

Deep Thinking: Where Machine Intelligence Ends and Human Creativity Begins
de Garry Kasparov

Human + Machine: Reimagining Work in the Age of AI

by Paul R. Daugherty , H. James Wilson

Working with Emotional Intelligence
by Daniel Goleman (Author)

The Impact of the EU's New Data Protection Regulation on AI
By Nick Wallace and Daniel Castro

Start with Why
by Simon Sinek

ZERO TO ONE
by Peter Thiel

The Art of the start 2.0
by Guy Kawasaki

Mastering the VC Game: A Venture Capital Insider Reveals How to Get from Start-up to IPO on Your Terms
by Jeff Bussgang

THE ENTREPRENEURIAL BIBLE TO VENTURE CAPITAL: Inside Secrets from the Leaders in the Startup Game
by Andrew Romans

REGULATION (EU) 2016/679 OF THE EURO-PEAN PARLIAMENT AND OF THE COUNCIL

of 27 April 2016 on the protection of natural persons with regard to the processing of personal data and on the free movement of such data, and repealing Directive 95/46/EC (General Data Protection Regulation)

https://motherboard.vice.com
https://www.wired.com
https://www.scientificamerican.com/
http://www.users.csbsju.edu/~eknuth/pascal.html
http://ai.stanford.edu/
https://deepmind.com/research/
https://www.etymonline.com/word
https://www.forbes.com
https://bbc.com/news/technology
https://www.brainyquote.com
http://wikipedia.com
https://www.boes.org/docs2/mking01.html
http://www.mgnet.org/~douglas/Classes/cs521/arch/ComputerArch2005.pdf
https://www.cyborgdaily.com/deep-learning-and-the-canadian-mafia/
https://www.securityweek.com/

https://www.recode.net/2015/7/15/11614684/ai-conspiracy-the-scientists-behind-deep-learning
https://www.phishtank.com/what_is_phishing.php
https://mbamci.com/les-batx-plus-forts-que-les-gafa/
http://www.homelandsecuritynewswire.com/darpa-working-major-cyber-security-break-through
https://hbr.org/1998/11/how-venture-capital-works

Distilling the knowledge in a neural network. (2014), G. Hinton et al. [pdf]

Deep neural networks are easily fooled: High confidence predictions for unrecognizable images (2015), A. Nguyen et al. [pdf]

How transferable are features in deep neural networks? (2014), J. Yosinski et al. [pdf]

CNN features off-the-Shelf: An astounding baseline for recognition (2014), A. Razavian et al. [pdf]

Visualizing and understanding convolutional networks (2014), M. Zeiler and R. Fergus [pdf]

Training very deep networks (2015), R. Srivastava et al.

Delving deep into rectifiers: Surpassing human-level performance on imagenet classification (2015), K. He et al. [pdf]

Dropout: A simple way to prevent neural networks from overfitting (2014), N. Srivastava et al.

DRAW: A recurrent neural network for image generation (2015), K. Gregor et al. [pdf]

Generative adversarial nets (2014), I. Goodfellow et al.

Rethinking the inception architecture for computer vision (2016), C. Szegedy et al. [pdf]

Inception-v4, inception-resnet and the impact of residual connections on learning (2016), C. Szegedy et al. [pdf]

Identity Mappings in Deep Residual Networks (2016), K. He et al.

ImageNet classification with deep convolutional neural networks (2012), A. Krizhevsky et al.

Rich feature hierarchies for accurate object detection and semantic segmentation (2014), R. Girshick et al. [pdf]

Spatial pyramid pooling in deep convolutional networks for visual recognition (2014), K. He et al.

Learning to learn by gradient descent by gradient descent (2016), M. Andrychowicz et al. [pdf]

Domain-adversarial training of neural networks (2016), Y. Ganin et al. [pdf]

WaveNet: A Generative Model for Raw Audio (2016), A. Oord et al. [pdf] [web]

Colorful image colorization (2016), R. Zhang et al. [pdf]

A Knowledge-Grounded Neural Conversation Model (2017), Marjan Ghazvininejad et al. [pdf]

Accurate, Large Minibatch SGD:Training ImageNet in 1 Hour (2017), Priya Goyal et al. [pdf]

TACOTRON: Towards end-to-end speech synthesis (2017), Y. Wang et al. [pdf]

ixelNet: Representation of the pixels, by the pixels, and for the pixels (2017), A. Bansal et al. [pdf]

Batch renormalization: Towards reducing minibatch dependence in batch-normalized models (2017), S. Ioffe. [pdf]

Wasserstein GAN (2017), M. Arjovsky et al. [pdf]

Ease.ml: Towards Multi-tenant Resource Sharing for Machine Learning Workloads VLDB 2018

Hyperband: A Novel Bandit-Based Approach to Hyperparameter Optimization Journal of Machine Learning Research 18 (2018)

Parameter Hub: a Rack-Scale Parameter Server for Distributed Deep Neural Network Training SoCC 2018

AI2: Safety and Robustness Certification of Neural Networks with Abstract Interpretation SP 2018

Bandana: Using Non-volatile Memory for Storing Deep Learning Models SysML 2019

Machine Learning at Facebook: Understanding Inference at the Edge HPCA 2019

ACKNOWLEDGMENTS

I would like to express my gratitude to awesome and beautiful J. You found the best way to bolster motivational energy that has pushed this project to fruition. You are making the world a better place.

ABOUT THE AUTHOR

Technology Guru and Entrepreneur. Actively consulting Fortune 500 companies on the next waves of Digital Transformations.

DETAILED TABLE OF CONTENTS